D0946451

"Marvelous! Packed with fascinating facts about the human animal. To be savored for the data and ideas alone, but watch out—you may dress differently, sleep better, remember more, view gender as a national resource and improve your efficiency. Weiner and Brown have explored human biology in ways that will start the office humming."

—Dr. Helen Fisher
Anthropologist,
American Museum of Natural History and
author of *Anatomy of Love: The Natural History of Monogamy, Adultery and Divorce*

"In this world of doublespeak, how wonderful it is to hear the clear, concise voice of Edith Weiner and Arnold Brown presenting a blueprint of human experiences to help each of us understand the physiological and psychological facts of work as we strive to make our mark in success."

—Annette Green
President,
The Fragrance Foundation

"Weiner, Edrich, Brown is, in my opinion, *the* pre-eminent firm in the area of Environmental Scanning. Their grasp of the complex dynamics driving change in the business environment is unparalleled anywhere. Perhaps more importantly, their ability to condense, translate, and communicate the implications of these trends, in terms their clients can understand *and use*, makes them one of the very few truly value-added suppliers of business intelligence in the market today."

—Brian Milton
Director of Government Relations,
Bell Canada

"I was particularly intrigued by the questions raised about the physiological and social impact of computers and communications technologies, speeding things up, creating a sense of 'timelessness,' and removing time as a pacing measure in the workplace. We need to think alot about *how* we work in the office of the future."

—Waring Partridge
Chief Strategic Officer,
AT&T Communications Products Group

OFFICE BIOLOGY

or

Why Tuesday Is Your Most Productive Day and Other Relevant Facts for Survival in the Workplace

Edith Weiner & Arnold Brown

MasterMedia Limited, New York

Published by MasterMedia Limited.

Library of Congress Cataloging-in-Publication Data

Weiner, Edith.

Office Biology : or why Tuesday is your most productive day and other relevant facts for survival in the workplace / Edith Weiner and Arnold Brown.

p. cm.

Includes index.

ISBN 0-942361-60-1 (cloth)

1. Industrial hygiene. 2. Occupational diseases. 3. Human engineering. 4. Stress management. I. Brown, Arnold. 1927– . II. Title.

RC963.3.W45 1993

813.6′2—dc20 93–9667

CIP

Designed by Jacqueline Schuman

Production services by Martin Cook Associates, Ltd., New York

Printed in the United States of America

10 9 8 7 6 5 4 3 2 1

DEDICATION

In memory of Ellie, without whom
this book would never have been.

CONTENTS

ACKNOWLEDGMENTS

A big chunk of our gratitude goes to Susan Strautberg, our publisher, for getting so excited about our doing this project that we had to see it through. We really appreciate her enthusiasm, patience and support.

To Anne Leon and Jim Perry, who typed the whole manuscript and then had to survive the wiping out of the disc that had it all in memory, a grateful nod. People who can read our handwriting and also enjoy what we write are worth their weight in gold. Well . . . certainly in a decent Christmas bonus.

And thanks to the hundreds of researchers—scientists and analysts—who gave so much of themselves to various pieces of the puzzle we have assembled. Their work is exciting and challenging, and we hope we have done them justice.

—Edith Weiner and Arnold Brown

INTRODUCTION

"Just WHO do you think you are?"

Do you remember hearing someone ask you that question when you were growing up? It was typically asked by an angry parent, but as you got older, it was also asked by teachers, snubbed friends, guidance counselors, testy bosses, jealous co-workers, disgruntled spouses, and, if you've now assumed the burden of parenting teenagers, your pouting offspring.

Think for a moment how we all answer that question. Mostly, we talk about our feelings. Our relationships. Our status. We hardly ever answer in terms of our physical being. Yet, emerging trends tell us that our answer may be in need of an overhaul.

In the past few decades, the marriage of computers and biophysics has led to astounding discoveries about you—the living being. From the increasing ability to scan your brain, to the rapid developments in mapping your genome (cracking your genetic code), it is increasingly possible to view the gradually unfolding instruction manual on what makes you tick.

Many ancient religions and healing practices were wrapped up in beliefs about how you, the human, are a unified body of mental and physical forces. *But*, in the seventeenth century, along came the great French mathematician and philosopher René Descartes, who said that humans were made up of two distinct and essentially separate parts—mind and body. This concept, known as "*Cartesian duality*," formed the basis of Western thought and science up through this century.

For a long time, Cartesian duality helped to advance modern medicine. But as we grew more dissatisfied with doctors who told us our pain was "all in your mind—take two aspirins and call me in the morning," we came to feel that there are a lot of problems with Cartesian duality. And our suspicions are being supported in research finding after finding. The more we learn scientifically about human biology and physiology, the less valid Descartes's split seems. We are having to see ourselves once again as an integrated whole rather than compartmentalized parts.

Brought down to the single human unit—*you*—there is emerging some of the most exciting evidence for how the physical and the mental are intertwined. Even as we write this book, there is a virtual explosion of information about what makes people happy, productive, anxious, compassionate, brilliant, depressed, or creative, and how these supposed "emotional" aspects of being are in fact linked to heredity, brains, and hormones, and how *all* of these interact with the physical environments in which you find yourself.

With sound waves and X rays, with computer-generated brain scans and electromagnetic imaging of the body, with heart monitors and broken genetic codes, scientists are closing in on more and more of the secrets of your physical being, becoming better able to discover just why *you* react as you do to anger, to improper lighting, to suppressed creativity, to different sounds and smells, to the challenges you meet with.

It was Charles Darwin, in his 1872 book, *The Expression of Emotions in Man and Animals*, who found some of the first evidence that people all over the world, in vastly different cultures, often have the same physical and emotional responses to the same experiences. For example, astonishment is almost universally expressed by the eyes and mouth being opened wide, and by raising the eyebrows.

About four decades later, one of the most important differences of opinion unfolded in the world of psychiatry. Carl Jung, a Swiss associate of Sigmund Freud, broke from Freud's total reliance on personal relationships (parents, siblings, "significant others") as a cause of emotional differences among people. In his most important work, *Psychology of the Unconscious*, written in 1912, Jung spoke of two parts of the unconscious mind—personal experiences and a common, *inherited* cultural experience. His theories of a *"collective" unconscious*, one that you and others might share based on your genetic makeup, was an idea that long existed in the shadow of the enormous popularity of Freud's theories. But, in recent years, the attack on Cartesian dualism has gained Jung (who died in 1961) and many other "holistic" psychiatrists additional credibility and followers at Freud's expense. Psychiatry and biology are now marching closer together toward insights into the real (biological) world of the mind—and they are on a march that could lead to some of the most startling and significant discoveries for centuries to come.*

Science now tells us that a hormone, oxytocin, may be closely linked to our feelings of love and infatuation. And how much we secrete of it may affect how nurturing or sociopathic we can be. In the lower part of the brain, one region called the amygdala may act as "command central" for many of our emotions and much of our memory, and how our amygdala goes, so may go some of our behavior. And you

*As an example, for many years Freudians claimed that the morning sickness many pregnant women have was mental—a physical manifestation of subconscious hatred of the husband or fear of the baby. Scientists today know that the senses of taste and smell become much more sensitive in pregnant women, because this enables them to protect their fetuses from toxins in food.

know that way you have of screwing up your nose and getting a headache from the smell of garlic? Well, your Aunt Martha did exactly that, and it's said she inherited the trait from her grandfather. How many of your reactions are in your genes? Probably many more than you know.

Did you know that hostility is bad for your heart? Did you know that anxiety changes the blood flow in your brain? That your stomach and intestines are laced with natural chemicals that react strongly to signals of nervousness from the brain? That lonely people are more at risk of illness? That depression affects the composition of your blood, thereby decreasing your body's ability to fight disease? That smells and sounds can trigger memories in the brain that can freeze you with fear or calm down your pulse rate?

Modern health care systems will eventually be greatly affected by the rapidly unfolding discoveries of our mind/body connections. Doctors are now being admonished to *know* the *person* in order to better diagnose the body. And the skyrocketing costs of health care will force all those interested—from businesses to insurance companies to the public—to take a holistic approach to health and wellness.

Our Bodies and the Office

It is our opinion that no environment offers more opportunity for controlling adult health care costs, boosting productivity, and enhancing our human spirit than the offices we work in day in and day out. No one has really taken a serious look at *all* the aspects of the work environment and how we interact with these from a mind/body perspective. Industrial engineering has offered us many research studies that help link aspects of the human body to productivity and safety. The new science of ergonomics is attempting to redesign the workplace to fit better with the workings of the human body. But much of this literature is not written for us, the workers, to understand and use. And most of it falls far short of fully integrating mind and body in the office, so that even those of you in management or human resources would not get a complete picture of what you could (should) be doing differently.

Clearly, the time seems right to move this discussion to where we can make use of it in our everyday lives.

An authors, we (Edith Weiner and Arnold Brown) do not have highly specialized backgrounds in biology or, for that matter, technology. (Although Weiner did earn a degree in industrial psychology—and Brown has spoken and written extensively on the impacts of technology in the workplace.)

We are strategic scanners, or what is more commonly called "futurists." Our work for more than twenty years has been in the area of tracking social, economic, political, and technological trends and gaining an understanding of where the world is going. We then translate that analysis into usable intelligence for our clients, most of which are corporations, or trade or professional associations. It is our job to help them identify the emerging trends and issues that will have an impact on their markets, their products, their images, their investments, their communities, and their employees. On a typical day, we might be assessing the future emergence of China, or the changing nature of religious beliefs, or the aging population, or the problems with water management, or the growing use of barter and countertrade in commercial dealings. We help organizations understand the impacts of new communications systems, political powers, illiterate youngsters, foreign investments, and new types of households. Our interests and analyses are far too numerous to catalog here. But there is one point we need to make, and make strongly. *One of the largest impact areas of all the trends we study is the workplace.*

Work is an important and guiding factor in the wealth and

well-being of individuals and communities. Issues that affect the workforce affect the economies of nations. And factors in the workplace affect, and are affected by, who and what we are.

So, we urge you to read this book because it could help you work smarter and happier, be more productive and creative, keep your health care costs down, and manage people and your work with more understanding of what you are doing, and why. We also think it will cause you to be better able to answer the next time someone asks, "Just *who* do you think you are?"

1 *The Sensory Jungle: Sights, Sounds, and Smells*

Look. Listen. Take a whiff. What do you really understand about that space you call your office? About the people you work with? It's time to pay more attention to what your senses are trying to tell you.

We all know how important our environment is to our well-being. But it means a lot more than just where the deer and antelope play. Your work space and co-workers make a difference in almost every aspect of your health, productivity, and performance. In this chapter, we'll look at how you interact with your workplace through your physical senses of sight, hearing, and smell.

Lighting

Sight is our most important sense: it provides about 85 percent of the information we receive from our surroundings. What we see depends on our eyes, of course, but also on the lighting. Lighting affects almost every aspect of our work: our comfort, efficiency, safety, and even mood. Our ability to perform most tasks depends, to a great extent, on the quantity and quality of the light. The Women's Occupational Health Resources Center recently surveyed nine hundred office workers, and found that they ranked lighting as the most important aspect of their work environment.

Studies show that productivity increases with proper lighting. This becomes even more important with the growing population of older workers, since eyesight declines with age. The retinas of a sixty-year-old person can only pick up one-third the light of those of a twenty-year-old. Older people also have less ability to discriminate contrast and colors, and they have less tolerance for glare.

To save energy in response to the OPEC oil embargo in 1973, the Social Security Administration reduced the lighting in its Baltimore office from 1,100 lux (a basic unit of illuminance) to 550 lux—and as little as 325 lux in some spaces. Productivity in one area where lighting had been reduced declined almost 30 percent, and employees began complaining of eyestrain and headaches. The original lighting levels were restored, and soon productivity returned to its original level. *The author of the study estimated that for every dollar saved by reducing lighting, $160 of productivity was lost.*

Light and Emotions

Changing the quantity, arrangement, and colors of lighting can have a profound effect on your moods and other emotional responses. One study had a conference room lit in six different ways. Subjects rated each different lighting on a range of factors including "friendly . . . hostile"; "pleasant . . . unpleasant"; "relaxed . . . tense"; "interesting . . . monotonous"; "clear . . . hazy"; "bright . . . dim"; "simple . . . complex"; "spacious . . . cramped"; and "formal . . . informal."

There were large differences in the responses to the six arrangements. Lighting used on both the table and the walls had the highest scores for qualities like friendly, pleasant, relaxed, interesting, sociable, likable, etc. But rooms with only overhead diffuse lighting got the lowest scores for these. Rooms with lighting only on the conference table were rated as small and cramped; while those with wall lighting were rated as larger and more spacious.

We need to know much more about how lighting affects our functions in the work environment. For instance, would we perform better at work if we could adjust the lighting at our own workplaces? Some studies suggest this is so.

Artificial Lighting Can Be Hazardous to Your Health

Virtually all of us have to work indoors, and that means under artificial light for most, if not all, hours of the day. Thus, we are deprived of natural sunlight. Why is natural light important? Because natural light, with the full spectrum of wavelengths including ultraviolet light, gives us a lot more than good eyesight or a sexy tan. It makes possible the most basic and critical life processes in our bodies.

After entering the eye, light energy reaches the hypothalamus, the part of our brain that coordinates most of our life-sustaining functions and maintains harmony within the body.

The hypothalamus, perhaps the most important part of the brain, determines our energy balance, fluid balance, growth and maturation, circulation and breathing, emotional balance, reproduction, heat regulation, and activity and sleep patterns. It directly stimulates and regulates our autonomic nervous system, which in turn regulates hormone secretions of the pituitary gland and the rest of our endocrine system, including the pineal, thyroid, parathyroid, thymus, adrenal, and pancreas glands and gonads. The endocrine system regulates the physical and chemical processes of human metabolism and the varying rates of chemical reactions in each of our cells.

The pineal gland, also located in the brain, responds to the natural light in our environment. It receives light-activated information from the hypothalamus to determine our daily biorhythms and rates of many bodily functions, including motor activity and sleep, body temperature, blood pressure, growth, tumor growth, the immune system, endocrine activity, reproduction, and possibly longevity and aging.

Without the pineal gland, we couldn't distinguish day from night or tell what season it is. It adjusts our entire physiology to the environment, and it only functions when stimulated by light through our eyes. One can imagine how many basic aspects of our health might be affected by the long-term deprivation of natural sunlight.

So what's wrong with the artificial lighting of our indoor workplaces (and homes, too)? It lacks the critical amount of ultraviolet light necessary for our health. Natural light lowers bodily stress, resulting in improved mood, behavior, and learning. We now know that nearly 25 million people in the United States are affected by "seasonal affective disorder" (SAD), and that it is caused by sunlight starvation. SAD causes changes in personality, including drastic mood swings, loss of energy, depression, weight gain, and often withdrawal (see chapter 2).

In addition, ultraviolet light activates the synthesis of vitamin D, necessary for the absorption of calcium and other minerals from our diet, lowers blood pressure, increases the efficiency of the heart, reduces atherosclerosis and cholesterol, assists in weight loss, increases the level of sex hormones, and helps treat diseases such as asthma, black lung disease, and tuberculosis. By using artificial light, we are getting out of touch with our natural environment and artificially manipulating and desensitizing the pineal gland's function—possibly throwing our moods and many body functions based on daily rhythms out of whack.

One of the most important studies on the influence of indoor environments on our health, productivity, and general well-being was done by Dr. Darell Boyd Harmon. He studied 160,000 schoolchildren to determine whether any factors in the classroom were contributing to any of their physical and/or psychological problems. He found that by the time the children graduated from elementary school, over one-half of them developed deficiencies related to the quantity of light available in the classroom.

With changes in lighting, seating, and decor, the children were observed six months later. There were substantial reductions in several physical problems including:

Problem	*Percent Reduction*
Visual difficulties	65
Nutritional problems	47.8
Chronic infections	43.3
Postural problems	25.6
Chronic fatigue	55.6

The study also found comparable improvements in academic achievement. Just imagine the enormous potential improvement in productivity many companies could achieve with similar changes in their work environments.

Dr. John Ott pioneered the study of the quality of light and its importance to the well-being of all living organisms. In the 1950s and 1960s, he measured the life span of laboratory animals kept under various fluorescent lights versus those living under natural unfiltered daylight.

The results were dramatic. Mice lived an average of 7.5 and 9.2 months respectively under pink fluorescents or daylight-white fluorescents, while those living under natural and unfiltered daylight were much healthier and lived an average of 16.1 months. Thus, he concluded that natural light is as important to the life and health of animals and humans as it is to plants.

Dr. Fritz Hollwich found in 1980 that limited-spectrum artificial lighting produces elevated levels of endocrine hormones ACTH and cortisone, which indicates high stress. This is one more reason you should make sure your office is lighted with full-spectrum lights, which have no such effect. Full-spectrum bulbs and natural light create much less stress on your nervous system than the standard cool-white fluorescent bulbs. Taken all together, the evidence points to a reduced number of sick days taken by subjects working under full-spectrum and natural lighting. So, you may actually boost your immune system simply by changing your light source.

Can Office Lighting Cause Building Sickness?

A British study suggests that office lighting may be one possible factor in producing a wide variety of work-related health complaints known collectively as "building sickness" or "sick building syndrome." The major symptoms include eye irritation, nasal blockage, running nose and sore throat, headache and lethargy.

Workers were surveyed in two high-rise office buildings. One had sealed windows and air-conditioning, and the other was naturally ventilated with open windows. Far more of the workers in the sealed building, which had smoked plastic screening on the windows, felt lethargic and complained of headaches. It was found that temperature, humidity, airflow, and airborne chemicals were the same in both buildings. The only difference was the amount of natural daylight—even though the sealed building had more artificial light, it actually had less illumination in the central work area than the naturally ventilated building.

You will read more about lighting in chapter 2, and more about "sick building syndrome" in chapter 8.

Color Dynamics

The National Association for Female Executives held a breakfast in 1991 for four hundred women who came to hear Dr.

Tessa Warschaw, a noted psychotherapist and author, talk about women and negotiating. She stood before the vast audience and said, "When I got dressed today to speak to you, did I wear black? No. Did I wear gray? No. I wore bright red. Why? Because I wanted to wake you up, to stimulate you, to have you listen to what I had to say."

A vast body of research supports her position. The color red has a physiological effect on the nervous system that stimulates blood pressure, respiration, and heartbeat. Pure blue, on the other hand, depresses blood pressure, heartbeat, and breathing. It has a soothing rather than exciting effect on us.

One of the most noted pioneers in the study of color, Dr. Max Luscher, pointed out that colors have a deep-seated psychological and physiological effect on us, which probably harks back to our primitive origins. When humans were at the mercy of the forces of night and day, darkness and light, the effect of dark blue was to suggest to the body that it was time to retire, to sleep. Daybreak, with its bright yellow, was the signal to the system to wake up, to energize. Attack was mainly in the form of hunting or fighting, and drawing blood, so that red came to signal attack and conquest. Self-preservation, the act of hiding or camouflaging in nature, or the idea of safety, was associated with green. Not only psychologically, but physically, our bodies respond.

Color vision has been shown to be related both to the "educated" part of the brain—the cortex—and the more primitive parts of the brain—the midbrain, as well as to the pituitary system. To distinguish color, identify it, name it, and react aesthetically to it, we use our cortex. But instinctive and reflexive responses to colors are dictated by their impacts on the midbrain, affecting our physical and glandular systems in ways we are only now beginning to understand.

Even if you are color-blind, it has been demonstrated that your body will react to darker colors by becoming more emotionally peaceful, and feeling less stress and tension. Brighter colors will spark energy and creativity, as well as aggressive or nervous behavior. Gray is, for everyone, neutral—it leads to no response.

So, to return to Dr. Warschaw's observation, if you want to make an impression, wear red. (Incidentally, rumor has it that red cars are more likely to be stopped by the police.)* And if you want to brainstorm new ideas or draw people out, decorate your office in bright, light colors. But if you need to concentrate, seek serenity, or put a client or employee at ease, think darker and more subdued.

There is a growing group of consultants who specialize in doing personal color analysis, so that you know which kinds of colors are best for you to wear because of your temperament and complexion, and the type of work you do. Overheard at a recent luncheon: "My dear, you should not be wearing that color. That's autumn, and you're a summer person."

We used to think our reactions to the colors we wear or the colors worn by others are purely aesthetic. Now we know that our body chemistry reacts along with our eyes, so there may be something to the idea of a more professional approach to the colors in your wardrobe.

Diane Ackerman explored the effects of colors in her wonderful book, *A Natural History of the Senses.* She noted that just as hospitals and schools use "passive" or "bubble-gum" pink to quiet children, children themselves paint in dark colors when they are sad, bright colors when happy. In a study done at the University of Texas, subjects watched colored lights as their hand-grip strength was tested. Red light caused them to grip 13.5 percent stronger. Elsewhere, hospi-

*Leatrice Eiseman, at Pantone's Color Institute, says this is not so. But she does suggest that if you paint your black Ferrari red, you'll get $10,000 more for it. The institute has a complete study of color preferences of thirty-five hundred consumers, and it includes clothing and furnishings. For information, you can call the institute at 201-935-5501.

tal patients with tremors who were exposed to blue lights were calmed, and the tremors diminished.

Manufacturers have long studied the effects of color on their products. Luscher pointed out that the physiological taste sensation associated with blue is sweetness; with green, it is astringence. Thus, sugar makers should never color their product green, unless they are going for a sweet-and-sour effect.

No doctor's office should be brightly colored, and most doctors instinctively know that. Their furniture and walls are visually more subdued, thus depressing any anxiety their patients may have. What signal do you suppose banks and insurance companies were sending to employees and customers when they almost universally decorated in gray?

The Effects of Noise: It's Not All in Your Head

Excedrin headache television commercials to the contrary, noise on the job is not just a minor irritation to be remedied by popping a pill. Noise can cause hearing loss, interfere with our communication and efficiency on the job, increase stress levels, and negatively affect our general well-being.

Sound Pollution

Audiologists (scientists who study sounds and hearing problems) are becoming increasingly alarmed about worker exposure to hazardous levels of noise.

Recent estimates indicate that between 7 and 10 million people work at sites where the level of noise (85 decibels or higher) presents a greatly increased risk of hearing loss. During 1978 to 1987, an estimated $835 million was paid in workers' compensation claims for occupationally induced hearing impairment, and it was estimated that between 8 and 11 million Americans suffered a permanent hearing disability. Most of this occurred in factory and blue collar work.

But office workers in major urban centers are affected by background noise in the streets and skies.*

And while many of us don't take notice of the din around us, whether caused by traffic jams or the whirring of machinery, the accumulated effect can take its toll. Hazards of noise pollution include high blood pressure and heart rates, altered blood chemistry, and nervous disorders. And new research points to the possibility that noise can even increase our vulnerability to toxic chemicals, by damaging centers of the brain that regulate metabolism and protect us from chemical effects.

In the 1970s, the U.S. Congress was very interested in noise abatement programs. OSHA (the Occupational Safety and Health Administration) and the EPA (Environmental Protection Agency) made some progress in controlling noise in industry and in setting standards in our environment. But most of the programs they began disappeared in the 1980s. As a result, many research activities aimed at developing quieter technology have gone by the boards.

The average urban American has come to tolerate a daily noise level of about 80 decibels, and over a forty-year worklife that takes a potentially dangerous toll on physical health and well-being.

If you are a speculator, invest in hearing aids. There's a good chance they will be as common as eyeglasses in the coming years.

*Seventy decibels is considered the maximum level for long-term safe exposure. Safety decreases substantially with every increase of 5 decibels. Consider the following noise sources: airplane (110–120 decibels), telephone (80 decibels), traffic (70–90 decibels), subway (80 decibels), noisy restaurant (70 decibels).

VDTs: More Than Meets the Eye

A number of office ills have been linked to video display terminals, and some of these will be covered throughout other chapters in this book. But, recently, a new problem came to our attention. Researchers at the University of Evansville have found that some VDTs give off high frequency tones that have been linked to user stress. In their study, women exposed to VDT tones complained of headaches, fatigue, and irritability, and they also experienced an 8 percent loss of productivity.

Why women? It seems that the tones generally emitted are at 16 kHz. Men can rarely hear above 15 kHz. Women can hear tones up to 18 kHz and report hearing noises half as loud as those detected by men. It's conceivable, though, that some men, too, might be affected by the VDT tones.

If you suspect your VDT is making noise that's driving you mad, but no one else around you can hear it, show them this page and then call in an audiologist or check out another VDT.

The Office Cocktail Party

How is it that, when at an office party or any social outing, we can be engaged in conversation with one or several people and focus only on the chatter in our group, despite the fact that many other conversations are taking place around us? The answer to this puzzle was published by a physicist, William MacLean, in his 1959 paper "On the Acoustics of Cocktail Parties." He pointed out that the sound you hear at any social gathering comes from two places. There's the direct sound coming from the person who's speaking to you, and the indirect sound that you hear only after it has been reflected off the walls, furniture, Italian leather shoes, and hors d'oeuvres.

Within each group, the person speaking can be heard above the background voices. For a while, this "signal-to-noise" ratio remains at a workable level. Until there are more speakers all around the room, and they start to raise their voices, as everyone gets involved in more and more animated conversation. At this point, we have trouble hearing our own conversation, so what do we do? We step closer to the person(s) we are speaking with. We do this subconsciously, but *biologically* we are compelled to. And that is because physics shows that the sound level reaching your ears from the person speaking to you varies in proportion to the square of the distance between you.

What does this mean in English? Only that if you move two times closer to the person you are talking to, his or her sound level will rise in your hearing fourfold. Three times closer, and your hearing of the speaker will increase ninefold. Therefore, it is the reduction of the distance between people, rather than louder speaking, that makes a difference in how well you hear your conversation. The more people at a gathering, the louder the background noise, the closer you will see people move together!

So if there's a reason for you to want to get close to someone at the office, make sure you get him or her to a well-attended gathering. Chances are you'll wind up close with no one feeling awkward at all.

The News About the Nose

One sense that has been generally undervalued but plays a large role in our daily lives is the sense of smell. It shows up in our language constantly—"I smell trouble," "This idea stinks," "Take time to smell the roses," "Let's sniff out the answers."

Certainly we know that we like to smell good (if not great), and we're repulsed by people or things that smell "bad." But we don't stop often enough to contemplate the true impact odors and fragrances have on our behavior, our mood, our productivity, and our decisions. There is a whole body of research developing on the biology of smell and the behavioral and emotional reactions of individuals and societies to different kinds of smells.

Subliminal Smells

There are certain odors that are perceived by our brains but that we are not aware of. Pheromones are one class of such chemicals that are given off by animals to communicate with others of their species, sometimes about food, territoriality, or dangers, but mostly about sex. Only recently has research been done that verifies that humans, too, may emit and react to pheromones. The search to identify human pheromones has been stepped up not only for its scientific importance but for the marketplace implications. There is a lot of money to be made by the companies that can bottle this stuff. It can potentially work to turn on a love interest, calm ruffled nerves, or elevate one's spirits.

Dr. David Berliner, a physician and venture capitalist who founded Erox Corp., is one of the pioneers in this field. Much of his research is based on the skin as the source of human pheromones. With the constant sloughing off of thousands of cells per hour, our skin could be a fabulous natural transmitter of chemical messages. But through centuries of ignoring such chemical messages, covering them with clothes and erasing them by bathing and use of deodorants and colognes, we have become far less conscious of the signals we are receiving. Yet, our behavior is being affected anyway in ways we don't realize. Berliner found that when he left extracts of his human skin substances in open vials in his lab, the friendliness and cheerfulness of the researchers increased dramatically.

Androstenol, a chemical that male pigs emit to attract female pigs, was used as a test substance by a group of researchers at the University of Southern California in 1987. Over the course of five weeks, two-inch square Plexiglas panels were taped to each of the four stalls in a men's washroom. On the first, third, and fifth weeks, plain alcohol was dripped onto these sheets. There was no impact on stall usage. During the second week, a solution of androstenol was deposited on the strip of one of the stalls. The use of that stall dropped significantly from about twenty visits per week to four for that week (and those four visits were made by the same person).

Androstenol is a very faint odor, hardly noticeable. And when a closely related chemical (androsterone) was dropped on that strip in week four, it had no impact on stall usage. The same experiment was run in the women's washroom, and it showed no such effect. So it seems androstenol gave a clear subconscious signal to men only that was territorial—a warning or discouragement from encroaching on the areas thus marked. But women are affected by different smells. Studies have shown that women living together or working closely together tend to see their menstrual cycles synchronized as a result of their bodies' subliminal response to each other's odors.

There is far more research that needs to be done in the area of subliminal or barely noticeable odors, but clearly our human biology is alert to signals in our work environment that we, ourselves, are not conscious of. Maybe that guy you didn't like is subliminally saying "Stay away!" to your nose. Maybe departments that don't get along are inhospitable to each other because of "marked" territories. And maybe the reason you want to work on the third floor is because, know it or not, you love the way it smells!

When Scents Make Cents

Dr. Eugene Grisante, chairman of International Flavors and Fragrances, one of the largest producers of flavors and scents for the perfume and food industries, believes that scent products of the future may be potent enough to make people eat less, work harder, drive safer, and sleep more soundly. These products may do many things we now use pharmaceuticals for.

Can office productivity be improved through the use of fragrance? Many researchers are "on the scent" of a new way of motivating people to be alert, to produce more, and to perform better on the job.

Kajima, the Tokyo construction giant, is in the forefront of those companies experimenting with and actually using fragrance delivery systems to affect mood and output in the office. The atrium of Kajima's new Tokyo headquarters is outfitted with a computerized duct system that can pump different aromas into the environment at different times of day. In the morning, workers are exposed to the smell of lemons, which invigorates them. Before noon, the scent of roses is pumped in, and a calming effect is produced, which enables the workers to enjoy their lunches better. In the afternoon, at about the time energy begins to flag, the woody odor of tree-trunk oils is used to pep up the workers whose productivity may be slackening. Shiseido, Japan's biggest cosmetics maker, plans to team up with Kajima to market similar environmental systems for office buildings, apartment houses, and sports complexes. *Kaori boomu* is the term for the fragrance boom in Japan, where even alarm clocks are designed to release the wake-up scent of a forest.

There are many other developments in the use of fragrance to affect mood and behavior. Researchers have found that people do better at jobs that require prolonged attention when they are exposed to puffs of certain perfumes. If you work at such a job, peppermint will stimulate you, while muguet (lily of the valley) will relax you.

Dr. Joel Warm, professor of psychology at the University of Cincinnati, has developed a plan for an air fragrance delivery system. He is working with a team to study the effects of odor on performance and stress, particularly with regard to sustained attention tasks. Dr. Warm is one of a number of fragrance researchers who have received grants from the Fragrance Research Fund, an organization formed by The Fragrance Foundation (an industry trade group) to specifically push forward the frontiers of knowledge about our sense of olfaction and the impacts of fragrances.

Annette Green, President of The Fragrance Foundation and coiner of the term "Aroma-Chology," is a leader in the exploration of the roles of fragrance in medicine, culture, interior design, personal health, and interpersonal relationships, as well as in the work environment. Grants awarded by the Olfactory Research Fund have led to some remarkable findings and applications. For example, Dr. William Redd, at Memorial Sloan-Kettering Hospital in New York City, has found that applespice has a particularly calming effect. He exposes his patients to it while they are undergoing CAT scans. This may save the thousands of dollars that are wasted when patients panic and cause their tests to be terminated.

Dr. Susan Schiffman, professor of medical psychology at Duke University, has uncovered some intriguing aspects of fragrance. One of her findings came from her counseling of couples in crisis: if one spouse does not like the odor of the other, there is not much chance of reconciliation. Dr. Schiffman points out that our reactions to many odors are highly individual, and that each person has his or her own unique fragrance. (As with a fingerprint, no two people have exactly the same smell.) Her research, along with that of others, leads us once again to conclude that we may get along well with, or be upset by, our co-workers and office associates based on a sense we are not even paying attention to—our sense of smell.

Robert Baron, of the Rensselaer Polytechnic Institute (Troy, New York), found that people he researched set higher goals for themselves when they were placed in a room periodically scented with air freshener. And they got along better with co-workers, becoming more likely to find friendly resolutions to conflicts. He found that better moods and greater willingness to take risks showed up on all the tasks he tested while subjects were exposed to fragrance. And wearing a discreet amount of scent at a job interview improved one's chances of being hired.

Researchers at RPI gave 120 student volunteers a scent test. Half were asked to perform certain tasks (such as proofreading and unscrambling letters to form words) in rooms without fragrance, and the other half worked in rooms sprayed with commercial air fresheners (apple cinnamon, light floral). Those exposed to the fragrance performed 25 percent better.

The Retail Planning Association in Seattle has been researching the impact of fragrance on customers in a sales situation. They have found that young women who shop at discount stores buy more when exposed to fresh natural fruit smells. Researchers at Chicago's Smell and Taste Treatment and Research Foundation showed that subconscious levels of certain odors persuaded 10 percent of shoppers to pay an extra $10 for a pair of sneakers. Several airlines and hotels are looking to aromatherapy (the therapeutic application of fragrances) to combat jet lag.

And so the research goes on. Mark Peltier, founder of Aromasys, in Minnesota, claims companies and universities have asked him to develop mood-altering fragrance systems. The states of mind he is most often asked to produce are alertness, relaxation, and refreshment. What do researchers like him recommend for each?

If you want to perk up at the office, try peppermint, lemon, eucalyptus, rosemary, or pine. If you need to calm down, try chamomile or lavender and clove, with floral and woodland notes. Do you need to be refreshed? Try a blend of citrus, pine and eucalyptus. Have a headache? Sniff peppermint. Tension? Bergamot. And if this is another day of palpitations and anxiety, surround yourself with ylang-ylang.

All Together Now

There is more and more talk of "environmentally responsive workstations," which would allow you to control the lighting, cooling, heating, airflow, and background noise in your work area. Rensselaer Polytechnic Institute undertook the first major study to prove that such a personal module would increase productivity and bolster mood and morale. When West Bend Mutual, a Wisconsin insurance company, moved into new facilities, workers who were given these personally controlled workstations showed productivity increases of 15 percent. When some of the controls were secretly and randomly disabled, productivity declined by 2 percent and some employees threatened to walk out if their machines weren't fixed.

Robert Baron of RPI sees an office of the future where people can control the total environment they work in. He includes fragrance controls, but says that it might be better to keep the fragrance levels low, perhaps even below the threshold of detection, so workers don't suspect they are being manipulated. Other researchers agree that if you feel you are being controlled, even by perfume, you are likely to find the aroma disagreeable.

Can't wait to get one of these total environment systems? The Atmosphere Manager, a $4,000 unit being marketed by Norm Automation of Sunnyvale, California, is the first commercially available setup we've read about. To create the right environment for working, negotiating, exercising, and even lovemaking and sleeping, this window unit lets you set the air temperature, humidity, amount and direction of airflow, lighting, fragrance, and even air pressure. We expect that more of such systems, at lower cost, will come onto the market. Clearly, it makes good sense to be good to your senses.

2 *Biorhythms: Working to the Beat of Night and Day, Dark and Light, and Season to Season*

Imagine that your body is a finely tuned instrument. When you play a well-timed, well-rehearsed tune on it, it responds with a rhythm so comfortable and clear that you can complete your entire performance without a hitch and to a standing ovation. But try playing something on it that you've never played before or that you've hardly ever rehearsed, and your timing is thrown off. The instrument that is your body can sound off-key, disquieting and disturbing. Now you find it hard to get on with your performance. You'd rather quit, you start to get irritable and frustrated, and the enthusiasm drains from your playing.

Like it or not, your body *is* an instrument that has been carefully crafted over the centuries of human development. Like flowers, birds, trees, fish, and even microorganisms, it has its own "biological clock"—a timing device that governs the daily rhythms of its function. Most of these occurrences are on a day/night cycle. Your heartbeat, blood pressure, temperature and respiration, and even urine excretion, experience heightened activity during periods of the day, and lessened activity at night.

Chronobiology and Circadian Rhythms

Day/night rhythms, or the twenty-four-hour cycle, are commonly called "circadian rhythms"—a term coined by Franz Halberg, M.D., of the University of Minnesota, from "circa" (about) and "dian" (day). In 1990, Judith Levine Willis, editor of *FDA Consumer* (a publication of the U.S. Food and Drug Administration), wrote a comprehensive article on biological rhythms, the study of which scientists call "chronobiology." Willis pointed out that the most well-known circadian rhythm is body temperature, which changes daily by one or two degrees in the healthy individual. Your body temperature generally peaks in late afternoon, and drops to its lowest daily point in the early morning. Pulse rate and blood pressure also reach their highest points in the late afternoon. This rhythm occurs even when you are bed-ridden for the entire twenty-four hours, and even if you alter the timing or amounts of your food intake. Willis also said that levels of glycogen, a carbohydrate essential to fuel the body, start decreasing about noon. By mid- to late afternoon, the body has used up much of its supply.

Studies have shown some remarkable things about the role of circadian rhythms in our lives. The senses of hearing, taste, and smell are more acute at certain times of the day. This cycle is related to our hormones—when our bodies release steroids, our sensory sharpness falls off. We are at our sensory peak at about 3:00 A.M. This declines rapidly to a sensory low at about 6:00 A.M., and then rises again to another peak between 5:00 and 7:00 P.M.

In addition, studies show that you will probably react more strongly to substances you're allergic to in the late morning, which is when antihistamine drugs have their greatest impact. If you are an average person, aspirin stays in your body longer if you take it at 7:00 A.M. than at 7:00 P.M. And if you have only one meal a day of two thousand calories, you will lose weight if you eat it for breakfast, but gain weight if you eat it for supper.

Researchers at Boston University have found that strokes are twice as likely to occur between 6:00 A.M. and noon than at other times of the day. This last bit of news comes from the Framingham Study, in which 5,070 residents of Framingham, Massachusetts, have had their health monitored since 1950. This study also revealed that hemorrhagic strokes—strokes caused by a broken blood vessel in the brain—strike most often on Mondays. Thus, Monday mornings really *are* the most dangerous part of our working life, and we probably have a deep-seated biological signaling device that urges us to ignore the alarm clock on Monday mornings. But we don't listen to it. We get up, get dressed, and go to work anyway.

So . . . Is Tuesday Our Most Productive Day?

Okay, so Mondays are clearly not days when we are perky, upbeat, and raring to go. On which day of the week, then, are our bodies primed for their best work performance?

We've all heard rumors to the effect that we should never buy a car manufactured on a Monday or a Friday. Even productivity professionals confirm this, but we spent many hours trying to track down why. Sue Miller and her husband run a management consulting firm called the Productivity Council, in Garland, Texas. The Millers have done a lot of research on workloads and workplace patterns, especially for banks. Sue tells us that "Wednesday is the 'slump' day." She describes the workweek this way: "Mondays we receive our work orders; Tuesdays we 'get in gear' to supply that demand; Wednesdays are a 'wind down' from the first part of the week; Thursdays we push to finish before the end of the week; and Fridays become like Wednesdays again."

Is this biology or is this cultural? Probably a good measure of both. People "feel" more or less productive on different days of the week, and their feelings tend to (but don't always) track with the Millers' findings.

The Gallup Organization did a 1991 survey for Accountants On Call (AOC) of full- and part-time American workers in which they asked, "Which day do you feel most productive?" and "Which day do you feel you are the least productive?" The largest percentage (39 percent) feel Monday is the *least* productive day of the week. Only 3 percent cite Tuesday. But when asked which day is their *most* productive, the highest proportion, about one in four, say Tuesday.

Most Productive		*Least Productive*	
Tuesday	24%	Monday	39%
Monday	21%	Friday	33%
Wednesday	17%	Wednesday	5%
Friday	13%	Thursday	5%
Thursday	9%	Tuesday	3%
Saturday	2%	Saturday	3%
Sunday	2%	Sunday	2%
Don't know	12%	Don't know	10%

So, while the other days come in for mixed reviews, Tuesday seems to stand out as the day most agree on as a high productivity day. But age plays a role. Younger adults (23 percent) actually cite Friday as their most productive workday, and almost one in three older adults were not able to pick a specific day as being more productive than any other!

Tuesday is far and away seen by the bulk of the Baby Boomers (thirty to thirty-nine years old) as their most productive day (31 percent cited it).

So maybe this has something to do with how you've spent your weekend. Or how you intend to. Or how, as you get older, you tend to pace your activity more evenly throughout the week. But since there aren't any cross-cultural studies on this, we can only speculate that people's perceptions of the rhythm of the week affect their bodies' willingness to attack the work with equal vim and vigor on a day-to-day basis.

Jet Lag

"Jet lag" is the term given to the total array of symptoms we get when we fly across time zones. Our bodies continue to operate on the rhythms we know, and it takes a while, even days, for us to reorient our body clocks. This is more true when traveling from west to east. When you are suffering from jet lag, you are sleepy in the daytime and can't sleep at night; you experience confusion, poor concentration, slowed reflexes, indigestion, hunger at odd hours, irritability, and even mild depression. So if your boss or co-worker has just come back from a long trip, that is not the time to ask for a favor, point out a mistake, or spring a surprise. And, if possible, you should avoid scheduling important meetings or events when you yourself may be traveling or returning from a long-distance journey.

Judith Willis offers these tips to those whose business or vacation plans do require jetting to and from faraway places:

- If staying in the new location more than two days, adopt local time for routines immediately upon arrival. If staying less than two days, maintain home schedules, if possible.
- If staying more than two days, several days before departure try to gradually shift sleeping and eating routines to coincide with time at destination.
- Before flight, avoid overeating and alcohol.
- In flight, drink water and juices, not alcohol. Don't smoke.
- If possible, break up long flights in one direction with layovers of at least a day.
- Allow plenty of time for sleep in the new location.

- After flying east, take walks outside in the morning to get used to earlier appearance of light. After traveling west, take walks outside in the afternoon to acclimate to later waning of light.

Melatonin and Light

While the factors of lightness and darkness are not all that feed into our circadian rhythms, light is a critical factor in the science of chronobiology (see chapter 1). Melatonin is a hormone that induces sleep and influences mood. We secrete it in the late evening from the pineal gland, which is located in the front of our brain, at the middle of our forehead. Some scientists think melatonin is responsible for sending information to our bodies that concerns the light-dark cycle that helps regulate our biorythms.

Recent research shows that production of melatonin (sometimes wryly called "the Dracula hormone") is directly related to light. Darkness stimulates production of melatonin, and light inhibits it. A study done at the Oregon Health Sciences University links this light-affected cycle of melatonin to regulation of the body's biological clock, keeping it attuned to external patterns of light and dark. Another recent study demonstrated that a small dose of melatonin in the morning will delay the biological clock—in effect, fooling the body into believing that it is still dark outside. Conversely, the melatonin dose will advance the biological clock if given at night.

What does the future hold? How about melatonin pills that can effectively alter our circadian rhythms. The pills are already in use—to help blind people adjust and regulate their biological clocks. Experts speculate that these pills can be used to reduce jet lag, decrease the discomforts of night workers, and effectively correct some sleep problems.

And what about our seasonal clocks? Why are some people, perhaps even several million Americans, seriously affected by the light and dark changes that occur from season to season? And why do the rest of us feel some symptoms of "winter blues"?

Seasonal Affective Disorder

SAD, or seasonal affective disorder (which was mentioned in chapter 1), has become a serious topic for scientific study over the past decade. No longer viewed as a minor and infrequent complaint, SAD has come to highlight the need for a more complete understanding of the natural cycles of the human body. It is clear that as winter rolls around and daylight contracts, many of us have the urge to sleep more and to eat more (especially carbohydrates). We get depressed and somewhat withdrawn. One researcher estimates that in northern regions, such as New England, 6 to 8 percent of residents suffer noticeably from SAD, and another 14 percent have milder forms of the condition. The incidence and severity increase the farther one lives from the equator.

Children are not immune. Youngsters who are sensitive to SAD also become irritable, bored, and easily upset. For these children, the winter months are not a time of glowing report cards. Their spring grades are clearly better. But most people who have SAD are in their late twenties and early thirties. Three-fourths of SAD sufferers are women. And SAD tends to run in families.

There are many theories as to the cause of SAD. Researchers have found that people with seasonal affective disorder have different biological clocks. In fact, they tend to secrete melatonin later in the evening (closer to midnight) than most people. There are other chemical imbalances in them as well. They have lower levels of seratonin, the substance within us that helps transmit nerve impulses, and they have unusual patterns of releasing dopamine, a key neurotransmitter.

The most interesting new research relates to the role the retina plays in our biorhythms. The retina is a part of the eye that is also part of a direct nerve pathway to the brain's clock mechanism. The retinas of people with seasonal affective disorder are significantly less sensitive to light, which is to say these people's eyes don't transmit light efficiently from the retina to the nervous system.

When bright light passes through the eyes of most animals, humans included, the brain stops producing melatonin. This is one of the principles behind hibernation of certain species. For people whose retinas are far less sensitive to light, the diminished light of winter, as well as dark and cloudy days at *any* time of year, can cause spells of serious fatigue and even depression.

The good news is that the syndrome can be "cured" with exposure to bright light. Dr. Raymond Lam, director of the seasonal mood disorders clinic at the University of British Columbia in Canada, was one of the first researchers to report on the link between reduced retinal sensitivity and SAD. Based on this work, Michael Terman (director of the Light Therapy Unit at the New York State Psychiatric Institute) and his associates have designed a set of computerized lights that resemble a summer dawn. SAD patients who use these in their bedrooms during the winter months are able to wake up earlier and have more energy. Terman believes it is in the early morning that our eyes are most sensitive to light and therefore that is the best time for treatment.

Studies on light therapy have demonstrated it can affect other problems with biological rhythms aside from SAD. It has been shown to relieve the sometimes debilitating depression of premenstrual syndrome (PMS). In fact, a 100-watt bulb one yard from a bed pillow has even alleviated irregular menstrual cycles. Scientists believe light therapy could be utilized to relieve many forms of chronic depression caused by the disturbance of our biorhythms, whether because of confinement indoors, jet lag, or night-shift work. Sleep disorder is a prime candidate for light therapy, as is overeating, because SAD often leads to eating binges in winter.

The more researchers uncover about the importance of light in our bodily rhythms, the more we are led to question the wisdom of windowless offices, dimly lit conference and meeting rooms, and predawn treks into the office to demonstrate our eagerness to get to work.

If you suspect that you or someone you know is having trouble at work as a result of the "winter blues," there are a number of books and organizations you might want to know about. Among them:

> The Society for Light Treatment and Biological Rhythms, P.O. Box 478, Wilsonville, Oregon 97070 (503-694-2404).
>
> *Seasons of the Mind: Why You Get the Winter Blues and What You Can Do About It*, Bantam Books, written by Dr. Norman Rosenthal, chief of out-patient studies at the clinical psychobiology branch of the National Institute of Mental Health. (Dr. Rosenthal is credited with having coined the term seasonal affective disorder—SAD—in the early 1980s.)
>
> *Seasonal Affective Disorder*, Harper, written by Angela Smyth. Smyth is a medical journalist and author who has written extensively for health and science magazines.

Summer Aggression

One of the more intriguing (some might say bizarre) studies reported on recently endeavors to link war with the seasonal cycle. Gabriel Schreiber, a biochemist and psychiatrist at the Beer-Sheva Mental Health Center of Ben Gurion University in Israel, says that his patients with emotional disorders are more aggressive during the longer days of the year. He extrapolates from this that national leaders, many of whom, according to Dr. Schreiber, "have psychiatric disorders," are more likely to begin wars in August than in December. (The opposite apparently holds true in the Southern Hemisphere.) Military people counter by saying that longer days and warmer weather offer better conditions for fighting. But Schreiber studied more than two thousand wars and found a strong correlation to the aggressive patterns of his psychiatric patients.

Sleep

About 35 percent of the American population complain about problems in falling or staying asleep. For one in every ten people the problem is chronic. Many of these people routinely suffer through days at the office displaying listlessness, fatigue, and irritability. Sleep is a critical part of our daily rhythmic pattern, and sleep disruption creates serious problems for our day at work. In fact, the National Sleep Foundation has pronounced insomnia more of a daytime problem than a nightime one.

Analysis of a study funded by the foundation and conducted by the Gallup Organization in 1991 shows that when compared to people without sleep problems, people with insomnia reported significantly impaired concentration, greatly decreased memory, a diminished ability to accomplish the tasks they need to perform, far less capability to deal with minor irritations, and a greatly reduced sense of enjoyment of relationships. The amount of the differences in all of these cases was quite startling. The impairment of on-the-job productivity is considered to be significant.

Despite the high incidence of this disorder, the foundation found that only 5 percent of insomniacs have ever specifically visited a physician to discuss their sleeping problem. Even during medical visits for other reasons, only three in ten insomniacs ever discussed their sleeping problems. Yet, clearly, they know they are suffering. Eighty-eight percent of people with no sleep problems often experience waking up in the morning feeling refreshed and ready to go. Only 53 percent of insomniacs (and 72 percent of those with occasional insomnia) feel as refreshed and ready when the morning rolls around.

How much sleep per day is important for your body rhythm to work well? It varies. There are people who can get all the rest their body needs on four to five hours of sleep per night. Some feel tired even after eight hours. Most people wake up refreshed and ready for a productive day at the office after seven to nine hours of sleep each night.

There are three types of sleep inadequacy, or insomnia:

- Transient insomnia can last one to several nights and is generally brought on by a change in sleep schedule, excitement, stress, or illness.
- Short-term insomnia lasts for several weeks and comes from the stress of extended medical or psychiatric illness.
- Long-term or chronic insomnia lasts longer, with sleep difficulty occurring nightly or frequently. Once again, physical or psychological stress is usually to blame.

Four in ten insomniacs report self-medicating with either over-the-counter medication or alcohol. The National Sleep Foundation clearly warns against this and advises seeing a doctor. If you have a sleep problem, they also suggest the following:

- Avoid caffeine within six hours of bedtime. Avoid alcohol and smoking one to two hours before bedtime.
- Exercise regularly. Get vigorous exercise in the late

afternoon. However, strenuous exercise before sleep may actually impair your ability to sleep.

- Don't nap.
- Establish relaxing presleep rituals, like a warm bath or a few minutes of reading.
- Go to bed only when you are sleepy, and use your bed for sleep only, not as an office or a place to watch television.
- If you are a bedtime "worrier," dedicate another time—perhaps thirty minutes after dinner—to writing down problems and possible solutions.
- If you can't sleep, don't stay in bed fretting. After ten or fifteen minutes, go to another room and read or watch television until you feel sleepy.

While the National Sleep Foundation cautions against napping, many of us find that our body clocks wind down in the early afternoon and we are drawn to want to take a nap. David Dinges, Ph.D., of the University of Pennsylvania in Philadelphia, believes that the "post-lunch dip" occurs regardless of when you eat lunch—or even if you skip it. When addressing a meeting of the Association of Professional Sleep Societies, Dr. Dinges said that afternoon drowsiness reflects a built-in human tendency to sleep in the middle of the day. People in siesta cultures are able to give in to the impulse, while those of us who live in nine-to-five societies are not allowed to indulge in this natural flow of our daily rhythm. "Given the opportunity," says Dinges, "more than half of all adults would probably nap."

If you do nap, or you occasionally get the chance to, you should know that it takes a while after you awake to rebound to your fully productive state. Dinges points out that for up to a half hour after sleeping many people experience mild grogginess, or "sleep inertia." Even if you feel as though you've had enough sleep, whether in the morning or after a nap, you won't do as well on an intelligence or reflex test as you will later on. Sleep, like many drugs, seems to have aftereffects that take time to wear off.

Ignoring Our Body Rhythm

There are many observers and researchers who believe that our culture is causing us more and more to operate in conflict with our body rhythm. Michael Young, in his book *The Metronomic Society: Natural Rhythms and Human Timetables,* says that we have adapted to a linear artificial tempo rather than a natural rhythm. Time has become a seriously valued commodity. Managers become victims, harassed at work and when they are away from work, all the while trying to work (and pay) to save time. He considers artificial lighting to have transformed our lives by creating a new society of night.

Activist Jeremy Rifkin, who vigorously opposes what he thinks are the adverse effects of science and technology on people, concurs. He says that in a society in which faster is always better, we have gone beyond a sustainable time orientation. We once learned to produce and consume at the same rate that nature could recycle and replenish. We now operate at the level of nanoseconds, which we cannot even perceive, let alone experience. (A nanosecond is a billionth of a second.)

More and more of us are caught up in time shifting and time squeezing. More than one out of six working mothers with children under fourteen holds an evening or night job, or a rotating shift. The same is true for one out of five working fathers. Many of us get in to work early and stay at the office until well after sunset most of the year. We are increasingly losing touch with the natural cycles of night and day and seasons, natural cycles that influence our bodies and our productivity as much as they do the planting and harvesting of crops.

As we finish writing this book, we continue to see new and important studies being done in this area. On January 11, 1993, *The Wall Street Journal* reported that Canadian researchers have found a higher rate of miscarriages for women who regularly work evening or night shifts.

It is clearly necessary for all of us, workers and management alike, to pay attention to our body clocks: to organize work around our most productive times, to give a nod of recognition to our least productive times, and to learn and try to accommodate to the clocks of the people we work closely with. We know now that we have some tools to help correct the imbalances in our rhythms. And occasionally we need to slow down and reflect on the productivity toll it takes on us as workers and as a society if we don't learn to care about the importance of biorhythms.

3 | *Gender: Bridging Brains, Babies, Bodies, and Brawn*

In the average workplace, stereotypes abound about the differences between men and women—how they "play the game," which are better "team" players, which are more focused on problem solving, which are the better "people" managers, which are the better number crunchers. Differences in opinions run the gamut from issues of sexual harassment to questions of total commitment to the organization. In her 1991 best-seller *You Just Don't Understand,* Deborah Tannen did an analysis of the different ways males and females communicate and relate to each other. Tannen explored how culture—sociological and environmental differences in upbringing and expectations—helps to divide the thought, communications, and motivational processes of boys and girls, men and women.

Since our book is based on biology rather than sociology, we were curious to see if we would find whether Tannen's theory about basic differences could, in fact, find merit on biological grounds. We did. There is mounting research to suggest that men and women are not only outwardly different, and socially different, and reproductively different, but that *Jane's brain is structurally different from John's brain.*

True for most mammals studied (e.g., rodents, monkeys, and birds as well as humans), brain differences between the sexes are apparently commanding more attention, controversy, and speculation. And more than any other chapter in this book, we suspect that the manner in which we highlight gender biology in the office will cause amusement, ruffled feathers, denial, enlightenment, and ample subjects for cocktail party conversation.

The Brain

The fun begins when we consider brain size. *Science* magazine reports that, on average, men's brains are about 15 percent larger than women's. This cannot be explained by the difference in body size alone (men are, on average, about 7 percent larger than women). And it is not explained by differences in intelligence (as we shall see later, men and women develop and display different patterns of intelligence).

In fact, one portion of the brain, called the SDN (sexually dimorphic nucleus)—a cubic centimeter of tissue that is highly sensitive to testosterone and estrogen, the male and female sex hormones—is 2.5 times larger in men than in women. From animal studies, it looks as though injecting testosterone in females causes their SDN to grow as big as males'; and by depriving males of testosterone, their SDNs shrink to the size of a female's. All this takes place in the part of the brain called the hypothalamus, and has much more to do with putting sex into overdrive than solving business problems.

Far more important is the corpus callosum, a bundle of nerve fibers that connects the brain's left and right hemispheres. Sandra J. Witelson, a behavioral neuroscientist at McMaster University and one of the most prominent researchers in the area of brain gender differences, has studied the corpus callosum extensively. She reports the finding that, while this brain region as a whole is larger in men than in women, one segment—the isthmus—is actually larger in women. The isthmus is the narrowest portion of the brain stem at the point where the middle of the brain and the back of the brain meet. Witelson has further noted that while the entire corpus collosum starts out larger in men, it decreases in size with age, whereas in women it stays the same size at least until the age of seventy.

Sex differences in the structure and size of the corpus collosum, and the isthmus within it, could be one major contributor to the differences found in patterns of thinking and perception between males and females.

In his book *The Creative Brain*, Ned Herrmann (former manager of management development at General Electric who went on to found The Whole Brain Corp.) reported on insights gained from a number of medical observations. The cumulative work he reviewed on the corpus callosum led him to write that:

- As many as 20 million more axonic nerve fibers are present in women, and, therefore,
- Impulses from a neuron in one hemisphere travel to its mirror-image neuron in the other hemisphere 5 to 10 percent faster in females than in males; thus, females can move ideas "back and forth" faster than men.

Herrmann's conclusions are supported by observations made in autopsies, after some forms of brain surgery, and by stroke recovery rates and patterns between men and women. For example, it is believed that women recover faster and more fully after a stroke because the nondamaged parts of their brain make up for the damaged functions. Further research is going on in this area, and if proven true, this may be a direct result of their brains being more synthesized and generalized, as opposed to specialized, as are the males'.

There have been other differences found in the structure of male and female brains, but it seems that the interconnectedness of women's brains is a major key in distinguishing their thought, perceptual, and problem-solving processes from those of men.

Command and Control

Perhaps the most important book to have been written on the subject is *Brain Sex*, by British authors Anne Moir (a Ph.D. in genetics) and David Jessel (a television writer and producer). Separately and together, they have painstakingly compiled the research on brain gender differences for almost two decades. And they did, and continue to, fight against the political unpopularity of much of the work they report on. For to accept the sum total of their research is to refute all beliefs that men and women could be made the same—that women could think, act, achieve, be motivated by, and be successful at all the same things and in all the same ways as men, if only they were socialized to do so. All the brain biology research to date concludes that men are, fundamentally, different from women.*

Brain sex differentiation begins in the womb. Both male and female fetuses are exposed to male and female hormones, but on the average, the boy embryo is exposed to a much larger dose of male hormone about six weeks after conception, at

*There are a number of professionals who do not accept the brain difference findings or theories that we will discuss in this chapter. Psychologist Janet Hyde, of the University of Wisconsin, believes the gender gap in test scores is narrowing, refuting the brain difference research. And social psychologist Carol Tavris, author of *The Mismeasure of Woman*, also suspects the motives, methods, and conclusions of many of the medical findings. After reviewing their analyses and the body of scientific work, we feel the science is more compelling. But we know that this is a thorny, controversial, and difficult issue for many, including a number of those whom we count as good friends. So we are notifying readers of a contrarian view in the event any would like to do more research on the objections to the work we report.

the critical time when his brain is beginning to take shape.* Another large surge of testosterone takes place at adolescence, when his sexuality comes on stream.

Therefore, it is not our genes directly that give us a male or female brain, but the degree to which our embryonic brains are exposed to male hormones. While the brain is developing, in the fetus, hormones dictate the neural pattern that is devised. Then, at puberty, the hormones wash over the brain a second time to switch on the pattern they primed it for.

And hormones continue, throughout life, to impact on the brain and various functions associated with cognitive skills. For example, Dr. Doreen Kimura, psychology professor at the University of Western Ontario, has found that women perform better on tasks involving verbal skill or muscular coordination when estrogen levels are high (just before ovulation in the monthly menstrual cycle). By contrast, women are better at tasks involving spatial relationships when estrogen levels are low (at the beginning of the menstrual cycle). She speculated that men, too, might experience fluctuations in cognitive skills along with daily changes in their testosterone levels (which are higher in the morning than in the evening). And Kimura's recent studies have confirmed that men have an annual hormone cycle, much like women's monthly cycle. She found that men performed better on standardized tests during the spring than during the fall. She believes this could

*An important but unpopular sidelight of the research is that, as with the development of the rest of the body, things at this stage could go wrong. A male fetus may have sufficient male hormones to trigger the development of male sex organs, but not enough to go on to prod the brain into a male pattern. A boy born with such a brain will "stay" female, although he will outwardly display a male body. Alternately, a female fetus could be accidentally overexposed, for one of a number of reasons, to male hormones and end up with more of a male thought pattern. Because the controversial nature of this research and the conclusions drawn are not what we are focusing on in this book, we cannot devote the necessary time and space to this issue. Suffice it to say that it may be biologically demonstrable that we really have a range of sexuality, with "male" and "female" as we most often interpret these being only two major clusters on a spectrum of human sexual diversity. We urge those interested in this to read *Brain Sex* and to track down some of the relevant research cited.

have especially significant implications for men in "spatially" demanding and high-risk jobs, such as surgeons and aviators.

Function Follows Form

The left side of the brain is the half that controls verbal abilities and the processing of information—logical, sequential thought processes. Speech, writing, and reading are all largely controlled by the left side of the brain.

The right side of the brain is where visual information is coded. This side deals with spatial relations, including sense of direction, shapes, and patterns.

Evidence points to the conclusion that the two sides of the brain are more coordinated in the female, less in the male. But the male brain, being more specialized in its functions, develops a higher level of spatial ability. Being able to picture shape, position, geography, size, and perspective in the mind's eye is a skill in which men clearly dominate, a conclusion confirmed by hundreds of scientific studies. Only a quarter of all women can perform assembly tasks on three-dimensional apparatuses better than the average male. And at the top end of the skill scale, men will outnumber women two to one. Throughout school, but especially from adolescence onward (when the second male hormone wash prepares the brain for maleness), boys will generally outperform girls in mathematics involving abstract concepts and theory. And the very best boys will eclipse the very best girls. Dr. Julian Stanley (Johns Hopkins University) and Dr. Camilla Benbow (Iowa State University) have studied highly gifted students of both sexes for two decades, and have found that not only did the best girl never outperform the best boy, but at the level of mathematical brilliance, for every exceptional girl there were thirteen exceptional boys.*

*We have no doubt that sociocultural, educational and economic biases are operating to make this ratio as large as it is. But what we don't know, and won't know, is just how

Boys also perform better at the hand-eye coordination needed for ball sports, once again demonstrating their ability to imagine, alter, and rotate an object in their mind's eye. They also tend to be superior to women in map reading.

Have you ever taken a drive with your friends or co-workers? You've probably noticed that men and women orient themselves differently, affecting navigational skills. Females tend to rely on specific landmarks, whereas men tend to rely on a "more primitive" sense of motion using remembered vectors. Dr. Thomas Bever, a psychology and linguistics professor at the University of Rochester, is an expert on how biological variables, like sex or handedness, shed light on brain organization and behavior. He found that, despite some overlap, most men and women use fundamentally different orientation strategies. That may be why men have trouble asking for directions; they don't have a sense that they are lost.*

Moir and Jessel report on numerous other research conclusions:

- Girls do better than boys on tests of verbal ability. (This evens out in adulthood.)
- Females are equipped to receive a wider range of sensory information, to connect and relate that information with greater facility, to place a primacy on personal relationships, and to communicate. Cultural influences may reinforce these traits, but the differences are innate.
- Girls and women hear better than men. Women show greater sensitivity to sound. Six times as many girls as

much smaller the ratio can become if these factors were not involved. The scientific evidence leads us to believe that perhaps we could change the ratio to even better than ten to one, rather than thirteen to one, but we would never, short of biological tinkering, approach the one to one level.

*Dr. Christina Williams, an associate professor of psychology at Barnard, was one of the first to discover that sex hormones played a part in male and female orienting strategies, by working with rats and mazes. Dr. Melissa Hines, assistant professor of psychiatry and biobehavioral medicine at UCLA, is one of those who see this as directly linked to the differential exposure of males and females to sex hormones before birth.

boys can sing in tune. Changes in volume of noise are more noticeable to women, which is why they are often more adept at discerning a changed "tone of voice."

• Women see better in the dark. They are more sensitive to the red end of the spectrum, and can see more shades of red than men can. They have better visual memory. Men, on the other hand, see better than women in bright light.

• Men tend to exhibit slight tunnel vision. They see a narrower field, with greater concentration on depth. Women "take in the bigger picture"—their wider peripheral vision is attributed to their greater number of receptor rods and cones in the retina.

• Women react faster and more acutely to pain, but their resistance to longer-term discomfort is greater than men's. Females are "overwhelmingly" more sensitive to pressure on skin anywhere on their body. Women have a tactile sensitivity that is so much more marked than men's that in some tests, the least sensitive woman is more sensitive than the most sensitive man.

• Women's sensoral superiority, clinically proven, is what may account for their being seen as having "supernatural intuition."

• Women learn better through verbal communication, men through visual.

Moving into the Workplace

Moir and Jessel take pains to show that even when social engineers are successful at eliminating gender stereotypes (as on the Israeli kibbutzim), children eventually sort themselves out by sex in areas of study and human interaction. Over and over, data confirm that "boys and men live in a world of things and space, girls and women in a world of people and relationships." And power in professional and corporate life has been sought and won overwhelmingly by males. The great majority of nurses are women, but almost all those who run hospitals are men. The great majority of teachers are women, but most principals are men. The management levels of corporations are now evenly populated by men and women, but women hold very few of the senior executive and board slots. The question keeps getting raised over and over—why is there so little parity in the ranks of the powerful? Is it only because women have also a home life they must pay attention to alongside their work commitments? Not likely. Many women are careerists, and do not have children or family distractions.*

Indeed, the management skills required now and increasingly in the future are people skills: motivation, persuasiveness, dispute resolution, articulation of visions and goals, etc. These are the very qualities women learn and use in raising children—often accommodating husbands at the same time. No business school teaches them, yet they are the keys to the success of any organization in this highly competitive world.

Is it then purely a matter of discrimination? No doubt this plays a significant role in the unequal allocation of power.

But could there be a third reason: a fundamental, biologically innate difference in cognitive approaches to power, success, and the measurements of these?

Corinne Hutt, a British pioneer in the study of sex differences, believes that women may not actually subscribe to the conventional definitions of success that men have adopted. She studied academics, a group in which men typically score higher than women in generally accepted definitions of success, such as the number of papers published. Hutt found that

*A 1988 study by Heidrick & Struggles, an executive recruiting firm, showed that one-third of female officers said their careers had affected their decision to have children (54 percent had none). A *Business Week* survey of top men and women in business found that half the women had never married or were divorced. And only half those married had children. About 95 percent of the men were married with children.

men were more concerned with academic prestige and institutional power, and women with developing students, fostering scholarship, and promoting institutional service.

In experiments, more men than women cite "achievement" as that which makes them happiest; women tend to merge success at work with personal achievement and the ability to make other people happy. Hierarchies and dominance are very important to the male—perhaps because his brain is structured to understand numerical standing and spatial placement. The politics of power, and the measurement of success by numbers—dollars, publications, titles—is much more meaningful to the male brain. Competition, single-mindedness, and comparison are all a part of a spatially dominant mind-set. For men, winning is a concept easily understood and a goal worth sacrificing a lot for, because it is a way of concretely measuring an outcome—success.

For women, on the other hand, it is not the short-term, immediate task that matters but the whole picture. A woman's cognitive world is much larger, and her broader perspective tends to put each individual numerical or quantitive event into a larger context. In the male mind, money is the symbol and acknowledgment of success. In the woman's mind, while money is as much *liked and wanted*, it is not the dominant confirmation of achievement.

This has been confirmed in a number of studies, the most recent of which was released in October 1992, by *New Woman* magazine and the National Association for Female Executives. In applying for cash awards in a new business contest, nearly one-third (759) of the contestants filled out a questionnaire related to why they wanted to start their own business:

The desire to feel passionate about their work (83 percent).
The wish for emotional independence (77 percent).
The wish for economic independence (76 percent).

The belief that they have a worthwhile product to market (76 percent).
The conviction that they are doing something of value for others (75 percent).
The realization that they cannot depend wholly on a man (64 percent).
Sexism and the glass ceiling (47 percent).
Wealth (22 percent).
Fame (20 percent).

The results clearly show that these women have a much higher regard for satisfying and worthwhile work than for wealth.

Not only do women tend to allocate themselves less salary than their male counterparts, they also tend to spend less on expense accounts, believing they have a responsibility to care for their employer's resources while involved in the company's business, just as they would care for their own if they were at home. A team at Rutgers University found that women researchers and laboratory workers generally rated their own performance lower than did their supervisors. Say Moir and Jessel, "Women, to this extent, discriminate against themselves in the comparative sexual earnings league."

There are women who, beyond question, are smarter than most men. And their heightened sensitivity and verbal skills could certainly make them excel in many male-dominated professions and careers, whether as doctors, ministers, congressional representatives, judges, bartenders, advertising executives, or strategic planners. But there are likely to be only two ways in which women will succeed in these fields in proportion to their ability. One would be for them to imitate men in almost every way possible, to "pass as one of the guys" in terms of competitiveness, aggressiveness, and ambition, while sacrificing interpersonal relationships, deeply felt values, and personal goals. Another way would be for the organization to change its culture—to reevaluate what achievement actually is, how it is measured, what makes for the success of the entire individual in his or her workplace and community. Want to take any bets on this second way happening in your lifetime?

Actually, that's not altogether fair. Many more women now own their own businesses than ever before (one estimate has it that there are more people now working for women-owned businesses than for the *Fortune* 500). So, even if you won't see massive culture changes in your large organization, you may witness it taking place in pockets of smaller companies.

It does also seem, however, that there are increasing numbers of women following the strategy of male mimicry. Is this because of their innate similarities to male-brain individuals? (Studies have linked some management women to childhoods as tomboys, and to mothers who took male hormones therapeutically during pregnancy.) Or could it be that the women who succeed in a male world know exactly how to model their behavior along the lines of men?

By suppressing the extremely important interpersonal and long-term strategizing skills of the female, male-dominated institutions continue to reward and compensate for only 50 percent of the important skills needed to make an organization succeed in the future.

And women continue to be frustrated by being told that the only reason they are not succeeding is because men are keeping them down. It may not be men, but men's rules, based on male thought patterns, that keep women "down." Because, biologically, they are tuned to different channels.

We spent several hours recently with Jayne Tear, president of the Jayne Tear Group, a consulting operation specializing in advising companies on "real" issues of gender difference, as opposed to "pop psychology." Jayne is committed to getting organizational talent to rise above stereotypes and obstacles,

and she finds brain biology to be one of the most promising areas for the future resolution of misunderstandings between men and women in the workplace. She helped translate the implications of brain differences into office reality for us.

According to Tear, the fact that women use both sides of the brain when engaging in verbal activities means they call into play a greater variety of sensory and memory data, making connections in various parts of their brains. Men, by contrast, engage only the left side of their brain in verbal activity; for them, speech and language are more highly specialized functions. This is one reason women seem to have more of a sense of intuition—their brains are feeding them a greater variety of information to tune into, synthesize, and analyze. But it is also why men tend to believe that women lack focus in their problem solving. Because they have a more interactive brain model when using language, women's brains may be likened to pinball machines, with thought traveling around and lots of lights and possibilities going off, making and remaking connections. Their own facial expressions, tones of voice, and physical gestures are called into play in the connective process. And so are their perceptions of all these characteristics displayed by others. *This is going on in the brain;* it is not just a socially inspired way of perceiving the world.

In men's verbal/language activity, only one portion of the brain is engaged in making up and hearing words, and the rest is "out to lunch." For them, often only the actual words spoken convey the message, and they will frequently complain, "She didn't hear me . . . this is exactly what I said, and that's exactly what I meant." So women will talk about their bosses or male co-workers as being insensitive, careless, angry, lazy communicators, who do not take into account how their communication is affecting others. Because in the woman's brain switchboard, it is not just what you say but also how you say it. While men are only aware of *what* they are saying—not their tone of voice, facial expressions, etc.—

women are picking up multiple cues in their mind's eye.

(With all this as backdrop, it could be concluded that the expanded use of electronic means of communication—e.g., computer networks instead of face-to-face meetings—will reinforce men's advantages. These electronic devices strip communication of all ancillary aspects; only the words matter.)

One kind of misunderstanding that results from these verbal/listening differences is fairly common in the workplace. A man listens only by using his ears. A woman listens with more of her whole body, and often will indicate that she is listening (paying attention) with facial expressions such as smiles or nods of her head and constant eye contact. Men, subconsciously, take this to mean agreement and fascination as opposed to politeness and attentiveness. Sometimes, this leads men to erroneously assume women agree with opposing positions or are defecting to a rival's camp or being won over by whoever may be speaking at the time.

Furthermore, women's language center patterns create a model in which speech has a rhythm, a fluctuation. When a thought is over, there is a pause, and that pause means an interjection by another party is appropriate. Men view this as an "interruptive" speech pattern, because they tend to speak without great fluctuation, and a pause is used for breath or thought.

Tear is convinced that, behind closed doors, each sex tends to see the other as poor team players and problem solvers. Without understanding how the two types of brains organize thought processes, fingers get pointed at personalities and gender stereotypes harking back to culture, upbringing, and office politics. The truth, however, may be purely biological.

According to men, women don't seem to be able to "focus" on getting directly from A to B. That's because, in the woman's brain, a lot of things are taken into account when considering how to get to B, including the relationships established with stakeholders along the way. This ensures that not only is B achieved, but it will last for a long time and lead to other profitable results not yet even imagined.

Thus, men may make good short-term goals achievers, and women better long-range planners. Yet, most long-term planning jobs are filled by men and are increasingly under the supervision of senior financial executives.

Similarly, men say women are not good team players because they're not likely to work well or at all with someone they don't like. This is often true. And while it may look like a lack of professionalism on the part of women, it could actually be the cognitive model of cooperation and camaraderie the female brain uses to organize and undertake work.

What Tear sincerely believes needs to happen in the office is for biology to be understood and used to the organization's best advantage. She advocates the equal or near equal employment of both men and women to solve problems and work on teams, because only then is the office working with a "full brain."

Clearly, were this to happen, not only would goals be met but problems would be solved better—from a different and perhaps longer-term perspective.

In looking for evidence that this might be the case, we turned our sights to the law profession—a male-dominated field in which more women are now present. Professor James Dobbs of Georgia State University used saliva tests (more easy and accurate than blood tests) to compare testosterone levels among lawyers. Trial lawyers had significantly higher levels of the male hormone than those who pursued more sedentary office jobs. And women lawyers had higher levels of male hormone than housewives.

And the growing number of women lawyers is beginning to lead to a revision of both legal theory and practice. From women's perspectives, legal problems and solutions are quite different from the ways men have defined them. Women's arguments are well known in cases of sexual harassment and discrimination, but their mark is also now being felt in contract, product liability, and copyright law. Where men had set structured, monetary, and rules-oriented legal theory, women lawyers are arguing intent, fairness, and caring as factors in judgments.

Women lawyers seek unstructured, innovative ways to solve business problems, believing contracts should be more flexible and agreements should focus on long-term relationships between signers. They believe that monetary judgment does not substitute for requiring personal retribution, and new kinds of judgments are needed if victims, businesses, and society are to benefit.

Studying this one profession is a clue to what we might expect as women enter all others: somewhat higher testosterone levels may lead women to greater risk taking and seeking out of male-oriented careers, but still being female means new orientations and approaches to the jobs that need doing.

The National Association for Female Executives (NAFE) and *New Woman* magazine give an annual cash award to three start-up women-owned businesses. In 1992, the three winners were companies run by women who were engineers, lawyers, and auto mechanics. The woman engineer and her partner were starting a business to educate about, and correct the overrepresentative exposure to, toxic substances in the environment of minorities. The lawyers had developed a program for teaching divorcing couples about the impact of their actions on their children. (And their two-hour course is now mandated by the Utah legislature for all divorcing couples with minor children.) The women mechanics have founded Mom's Garage, a place for women to take their cars and not be ripped off or talked down to. (In fact, they will teach women how to save money by taking care of their own cars, they will provide a child play area, and they will give free advice to women who are buying used cars. Furthermore, they hope to inspire more women mechanics—and there are quite a few—to own their own garages, and to train more women to become mechanics.)

So the point is not that there are women's jobs and men's jobs, but that those who pursue any career will do so using a uniquely patterned thought process that often reflects gender brain differences.

Pregnancy—a Two-Sex Story

There is one thing we know for sure about gender biology—women get pregnant, men don't. And that is a cause of serious discomfort and discrimination. Despite laws that make it illegal to discriminate on the basis of pregnancy, many women are still being let go, penalized by demotion or retracking, treated with condescension or resentment, and discouraged from returning to their jobs if they now have another focus: family.

Even in "family-friendly" companies, where male executives are offered paternity leave, the reality does not track with the credo. While about one-third of large U.S. companies offer paternity leave, hardly 1 percent of those eligible elect to use it. They are reluctant, in a man's world, to convey the image that they are less than 100 percent committed to their career and company. That's why not only are women leaving their corporate executive suites, but men are, too. The heat is getting turned up for both sexes. Childbirth and children are becoming more important for those who increasingly see themselves in an institutional setting that is more and more uncertain, punitive, stressful, and unrewarding.

An important sidebar to all of this is that there have been many efforts through the years to bar pregnant women from working in environments that pose potential risks to the unborn fetus—in some cases, barring all fertile women even if they are not, or say they will not become, pregnant. Despite the fact that a 1991 Supreme Court decision ruled employers may not bar women on this basis, there is still a great deal of controversy related to prenatal risk of exposure, not only on the factory floor but in the office environment. From low-level VDT radiation to indoor pollutants (see chapter 8), full information about risk does not remove the risk.

But that risk is also to men! Radiation and other substances that damage the genetic material in cells may even be more threatening to men than women, because sperm cells are continually dividing, whereas the eggs in the female ovaries are not.

Martin Gardner, an epidemiologist and medical statistician, in a 1990 study of England's Sellafield nuclear plant, concluded that a father's exposure to low-level radiation at work might increase his children's risk of leukemia. Of many variables studied to explain the high incidence of childhood leukemia in Seascale, England, the only strong correlating factor was the father's employment at the nuclear plant and his radiation dose *before his child's conception.*

Recent research at Boston University by Gladys Friedler bears this out. Her experiment with male mice led to her conclusion that fathers, not just mothers, can contribute to all sorts of congenital problems. She also cites other studies that have linked birth defects to the occupation or habits of the fathers, including those who were fire fighters, painters, garage workers, or even smokers.

So, for all those of you who believe that only the fertile or pregnant woman needs to be shielded from workplace risks, here is a new and important caution. Men are just as vulnerable to prenatal workplace exposure risks; they must also curb personal habits like consuming alcohol, smoking, or doing drugs if they are thinking about ever conceiving a child. This is especially important since men are twice as likely to abuse drugs and five times as likely to abuse alcohol as women. (However, this may be changing. More single, well-educated young women are now being treated for alcoholism than ever before.)

Take It to Heart

According to *Science News*, employed women have higher blood levels of high-density lipoprotein (HDL) cholesterol than their nonworking counterparts. HDL is the "good" cholesterol that lowers heart attack risk. Scientists and doctors had been warning women that by taking on all the trials and tribulations of work and career, they were also risking the higher rates of heart disease experienced by men. On the contrary, coronary heart disease mortality among women in most Western nations has fallen during the last decade.

There is, of course, the possibility that women who juggle a career and family have a higher activity level, which may cause the increase in HDL. Also, women who hold jobs, especially higher-paid managerial positions, may be better able to afford better medical care and nutrition. Working mothers know, however, that, despite what men believe, women can find work at the office a calming oasis in an otherwise frantic life. In fact, it might be argued that motherhood is a far better preparation for the hectic demands of organizational life than is an M.B.A.

Or maybe the National Opinion Research Center may have the answer to working women's healthy hearts. In a survey of three thousand women, they found working mothers have sex about twice a week, on average, compared with once a week for all wives. And about one in four working mothers admitted to having extramarital affairs, two-thirds saying these affairs were with men they work with. Whether because of unhappy marriages or wanting to have fun—two of the main reasons given—it's clear that office biology is more than just an issue of cholesterol.

By the way, for those of you women who are examined by

the office doctor, who is almost always a male, take heed: James McCubbin of the University of Kentucky College of Medicine has found something you should know. In a study of two hundred people, women showed markedly increased blood pressure readings when tested by a man. The level was finally stabilized to a lower measure by the fourth reading. Therefore, he believes that women may be at risk for being wrongly diagnosed with borderline hypertension, especially if the screener is a man. So, whether you're tested at work or on your own, if the doctor or screener is a man, ask for several pressure readings!

4 *The Body Burden: The Inescapable (?) Effects of Aging*

"Everybody wants to live a long life," said satirist Jonathan Swift, "but nobody wants to be old." In America, which has an almost obsessively youth-oriented culture, this is even more true than it was in England in Swift's day. We celebrate youthfulness and all the vigor and energy that go with it.

But we are living in a time when this culture is increasingly in conflict with demographic reality. The American population is aging; perhaps most important, the Baby Boom, that one-third of the population that has dominated the public agenda for the last three decades, is aging.

By 1990, more than 30 million Americans—over 12 percent of the population—were sixty-five or older. By 2020, with the Baby Boom aging, more than 20 percent of Americans will be sixty-five and up. (And as many as 7 million of them will be over eighty-five years old.)

In 1970, 42 percent of the workforce was in the sixteen–to–thirty-four age group, 40 percent was in the thirty-five–to–

fifty-four age group, and 18 percent was fifty-five and over. By the year 2000, according to the consensus of demographers, the numbers will be:

- Sixteen to thirty-four—38 percent.
- Thirty-five to fifty-four—51 percent.
- Fifty-five and over—11 percent.

The workforce, in other words, is becoming middle-aged.

As futurists Joseph Coates, Jennifer Jarratt, and John Mahaffie point out in their book, *Future Work*, there are some very important health-related implications of these demographic shifts. As we get older, we tend (at least most of us) to become more sedentary. And office work, unlike factory or farm work, is almost totally sedentary in nature. Coates, Jarrett, and Mahaffie estimate that we Americans, on average, spend 90 percent of our time indoors.

The convergence of these two tidal forces can have great impacts on work, workers, and the workplace. The sedentary life tends to promote or even speed up the natural declines that accompany aging, particularly the decline in the muscles. In addition, as we show in chapter 8, the interaction of this decline with the possibly unwholesome environment in today's office buildings would significantly increase the health problems associated with degeneration and aging.

Up and At 'Em

As for Baby Boomers, they seem unwilling to go quietly into dim and decrepit middle age. Many are maintaining or even increasing their exercise routines. Experts are just about unanimous in approving this as perhaps the single most important step people can take for themselves. Exercise, as has been proved many times, can delay, offset, or even reverse the ravages of aging. It can keep your heart and circulatory system vigorous well into old age; it can keep your lungs and respiratory system healthy; it can maintain muscle tone; and it can enable you to go on having feelings of well-being and vitality. (A problem with it, though, is that it can't necessarily keep the hairline from receding.)

Studies of astronauts have shown that some muscles can lose up to 40 percent of their mass in just a few weeks if circumstances prevent their regular use. Muscles adapt with remarkable speed and flexibility to the body's needs. If you don't use them, you won't have them.

Scientists are studying the possibility that in the future, we will be able to manipulate muscle fibers genetically. Perhaps DNA injections or alterations could program muscles to grow bigger and stronger—without exercise or steroid abuse.

But, for now, many employers are responding by providing exercise facilities or in other ways encouraging their people to get into and keep up programs of regular exercise. As more employees move into that thirty-five–to–fifty-four age group, the demand (and need) for this will increase.

Another way to keep decrepitude at bay is diet. And there, too, Baby Boomers are leading the way. They are listening more to food and nutrition authorities, and they are learning to eat not just to stay alive, but to stay healthy. For more details, see chapter 5.

Sensing the Years

Exercise and diet can help. But some aspects of aging seem, at least as far as we know now, to be inevitable no matter what we do. There are physical changes that are so universal that individual life-style seems to have, at best, marginal effects on them.

Vision is one. Sight is keenest in young adults, those in the twenty–to–twenty-five age group. After that, it begins to decline—gradually at first, but accelerating after age fifty-five or sixty. The average sixty-year-old person needs three times as much light to see an object as does a twenty-year-old. The ability of the eye to focus on things that are close is sharply reduced—a decline that becomes increasingly apparent to people once they reach their forties.

As you age, your eyes change. The lenses become thicker, and this tends to make glare more of a problem. So fluorescent lights can be troublesome. The ability to differentiate among shades of color diminishes; the lens, cornea, and fluid in the eye yellow as you get older, and this changes perceptions of colors and shades. Blue and green, for example, become more difficult to distinguish.

One glaring example of inattention to this phenomenon of aging eyes is the current fad for black TV and audio sets. Dark backgrounds make reading dials even harder for aging people. In some cases, eye exercises may retard or reverse some of these declines, but beware of extravagant promises. In most cases, glasses will still be needed. There is some evidence that vitamin A is helpful in slowing down eye aging. Bugs Bunny, after all these years, still sees exceptionally well. So if you like to snack, slice a carrot into long, thin pieces, let them sit in a cup of very cold water on your desk, and munch on them as a snack throughout the day. Good for the eyes and the waistline.

Hearing also declines. Over time, the ear loses its ability to detect higher frequencies—particularly those above 1,000 cycles per second (one kilohertz). The human voice can cover a spectrum ranging from under 100 cycles per second to more than 8,000. The ability to hear much of this begins to fade as people age. (Men, it should be noted, begin to experience appreciable hearing decline on average six years earlier than women. But, because women speak in higher frequencies, they are the ones who become more difficult for aging people to hear.) It is sobering to consider that the ability to detect speech well across the sound spectrum can be 5 percent less for people in their forties than for people in their twenties.

The sense of smell also declines with age. Other changes include declines in flexibility, reaching ability, and reaction time. As you will see in chapter 7, office ergonomics needs to focus more on what it means to have an aging workforce.

Given the current concern with accommodating the workplace to handicapped and disabled people, supported by the Americans with Disabilities Act, it is constructive to note that accommodating an aging workforce can be responsive to the needs of the disabled—and vice versa.

Memory also declines with age. But it may be that the problem is not memory so much as it is the ability to recall. That is, the information may still be there; it's just harder to find it. This theory seems to be proved by experiments that show that memory can be improved in older people—particularly by repetition and the use of mnemonic techniques. For example, if you are assigned to work with a client group and you are told the members' names are Roberta, John, Allen, Tawanda, and Bruce, you might arrange the sounds of their names in a sentence (nonsensical but memorizable) like: "John's Town Robe is All Brute." Remember the sentence and you can recall their names. Similarly, if you have several names or things to remember, you might arrange the first letters to spell a word, which will serve to remind you of the list. Let us say you need to order a *s*tapler, *p*encils, *t*ypewriter ribbon, and *e*nvelopes. You might just think of the word "step," each letter of which will remind you of what you need. These sorts of devices come in handy for everyone, and in fact many doctors will tell you that it is nearly impossible to get through medical school memorizing all the bones and muscles and parts of the body without such gimmicks. Memory fails all of us from time to time; aging only compounds the problem. Using memory tricks is nothing to be ashamed of.

Menopause

Perhaps one of the most significant developments resulting from the Baby Boom's refusal to accept passively the decline and decay of aging is that we now openly talk about things that formerly were ignored or whispered about. Case in point: menopause.

In one remarkable month—May 1992—menopause burst on the public consciousness like a bombshell. *Newsweek*, the Baby Boom's *Michelin Guide* to life and culture, made menopause its cover story for the May 25 issue. *The New York Times,* which likes to see itself as the official newspaper of America, had several feature articles on the subject in one week in May. And, also during May, two major books on the subject hit the bookstores: *The Silent Passage* by Gail Sheehy, a well-known chronicler of "hot" subjects; and *In Full Flower,* by Lois W. Banner, a historian who teaches at the University of Southern California.

Why, all of a sudden, is menopause such a hot—and such a public—topic? The answer, of course, is that the Baby Boomers are arriving, in their usual vast numbers, at that station in life.

As *Newsweek* points out right at the beginning of its feature article, more than 40 million American women—the cream of the Baby Boom—will experience menopause over the next twenty years. And, as *Newsweek* also points out, they are not "going silently." They are doing it with customary Baby Boom public clamor.

Previous generations bore menopause stoically. Women just didn't talk about it. As a consequence, not much was known. (One of the things not known about it was that men, too, may undergo a kind of menopause, as we will discuss later.) Now, however, women want to know; they are asking questions, and they are finding that answers are hard to come by.

Even though perhaps no more than 15 percent of women are troubled enough by symptoms of menopause to seek medical help, the fact is that all women go through the significant biological changes involved.

Foremost among these changes is a big drop in the production by the ovaries of the female hormone estrogen. Production of estrogen usually begins to decline around age thirty-five. The decline speeds up at menopause, when the ovaries, in effect, close up shop and menstrual cycles end or fade away.

This certainly could impact upon the power structure. Helen Fisher, a noted anthropologist at the American Museum of Natural History, sees the Baby Boom women as a massive force entering menopause, a time when estrogen levels decline, unmasking natural levels of testosterone. Their testosterone, clearly linked to aggressiveness and now less tempered by high estrogen levels, has led middle-aged women in many societies to gain greater influence in political, religious, economic, and social life. As we look forward to a bumper crop of mid-life women, we can look forward to a major power surge. (Witness the 1992 election results.)

Estrogen, we are discovering, affects much more than just the reproductive organs. Many other parts of the body have estrogen receptors, and the reduction in estrogen can have significant and often adverse effects on skin, hair, heart, bones, brain, etc.

One symptom of the estrogen decline is the phenomenon known as "hot flashes." The severity of these can vary greatly: some women experience only brief hot moments; others have lengthy and extremely uncomfortable hot periods, day and night, that can make them constantly tired and irritable, and completely throw off their workday. For some, these symptoms can go on for up to five years.

Other results of menopause, largely attributable to lack of estrogen, include vaginal problems, increased cholesterol levels, and bone loss. The bladder and urethra can also be adversely affected, and this can result in infections and other complications, including incontinence.

Many women face embarrassing episodes at work and suffer silently because, as yet, not enough has been done to discuss these issues out in the open.

To deal with all of this, more and more women are using hormone therapy to replace their lost estrogen. Estrogen therapy can limit or even reverse some of the damage resulting from menopause. According to *The New York Times,* women who undergo estrogen replacement suffer 40 percent fewer fractures than those who don't. Women who use estrogen replacement are less at risk for heart attacks, too—up to 50 percent less, according to a study done by Harvard Medical School and Boston's Brigham and Women's Hospital.

Unfortunately, there is also a risk to estrogen therapy. After five years, the risk of cancer of the endometrium (the lining of the uterus) increases very substantially. And although the link between estrogen replacement and breast cancer is not clearly established, there have been enough negative findings to cause concern. The greatest risk appears to be for women with a family history of breast cancer.

Some forms of therapy add another hormone—progestin, an artificial form of the hormone progesterone—in an attempt to counteract the carcinogenic potential of estrogen, particularly for the uterus. While this may work (it has yet to be officially approved), progestin does have some negative side effects, including weight gain, vaginal bleeding, and breast tenderness.

To date, less than 20 percent of menopausal women use hormone therapy. This may be due to health concerns, but it may also indicate that most women do not suffer extreme symptoms. There are women who use nonhormonal medications, such as phenobarbitol, to relieve their symptoms; others use vitamins and herbs, such as vitamin E (an antioxidant), vitamin B_6 and ginseng.

Some medical authorities feel that diet plays a significant role. Recent research shows that Japanese women, who traditionally have a low animal-fat diet, are less subject to such menopause symptoms as hot flashes than American and European women.

Some doctors, and some women, say that a good exercise program can help counter the effects of menopause, by preventing bone loss and reducing cholesterol. Others use and recommend biofeedback.

Regardless of what they do or how long it lasts, many women experience symptoms that significantly affect their work. They suffer from migraine headaches, irritability, forgetfulness, and stress. (Because estrogen is stored in fat cells, thin women may suffer more of these symptoms than their plumper counterparts.) Recognizing this, women need to make a special effort to ask friends and co-workers for honest feedback, and to be less sensitive when told they are out of sync with their traditional demeanor and performance. Sometimes only through such honest feedback from someone we trust can we attempt to consciously alter or compensate for shifts in productivity and mood that negatively affect those around us.

Male Menopause

Is there also a male version of menopause? Most scientists say no, primarily because production of testosterone, the male equivalent of estrogen, does not abruptly end. But there is evidence that the supply of testosterone, one of the androgens (male hormones) produced by the testes, does decline with age. This decline has been linked to some of the outward manifestations of advancing age: lower fertility, lowered interest in sex, less energy, more body fat, deteriorating muscle tone. Some doctors believe that the decline in testosterone may be due more to life-style factors—overweight, lack of exercise, etc.—than to aging. Some studies have found high testosterone levels in healthy men over fifty, although some others have found a decline.

As with women, hormonal decline in men affects moods. The drop in testosterone has been linked to lower energy levels, lowered ability to concentrate, and weakened aggressiveness.

Some men are using testosterone supplements, although the medical community does not agree on their value. According to many doctors, hormone therapy for men has yet to prove that it is either safe or effective. Meanwhile, there is agreement that—as with women—regular exercise and a proper diet can significantly alleviate many of the effects of aging.

In the meantime, as women come into a more testy time of their life, men are slowly beginning to mellow. So it may well be that in the office, the two sexes start to become more like each other after age fifty, and that could certainly have implications for interpersonal styles and relationships.

Life Extension

A big factor in the relationship between aging and work is the extension of life that has occurred in our time. We are living longer. The average life expectancy for all Americans in 1990 was 75.4 years. For white males, it was 72.6 years, and for white females, 79.3 years. For black males, the figure was 66 years, and it was 74.5 years for black females. The major contributors to less longevity for African–Americans are higher incidences of heart disease, strokes, and cancer.

The number-one cause of death for all Americans is heart disease, which is responsible for more than 700,000 deaths each year in the United States. Cancer is second, and it is increasing, while death by heart attack is decreasing. Since 1965, there has been a 25 percent reduction in deaths from cardiovascular diseases. Much of this positive development, apparently, is attributable to improvements in life-style—in particular, less cigarette smoking, more exercise, and better diet, and perhaps even stress reduction programs at or related to work.

In a report we prepared for a Society of Actuaries conference on the future, we pointed out that medical and technological advances in areas relevant to aging—neurobiology, gerontology, immunology, and genetics—seemed likely to affect profoundly both longevity and the process of aging itself. We will not only live longer, we will be younger longer.

Perhaps within twenty years, according to some experts in the field, we may be able not only to cure many of the afflictions of aging, we may also be able to slow down the aging process.

According to a study done in 1987 for the Alliance for Aging Research, scientific and medical advances that are in the offing include:

- Identifying the genes that control aging and longevity;
- Being able to alter those genes to regulate the aging process;
- Finding the genetic basis for diseases and afflictions associated with aging, such as Alzheimer's;
- Finding cures for heart disease and cancer;
- Successfully healing or even preventing such manifestations of deterioration as osteoporosis and osteoarthritis.

Dr. Jan Vijg, head of molecular biology at the Institute of Experimental Gerontology in the Netherlands, says that nobody really dies of old age. They die of age-related diseases. Once we conquer disease, not only will we live longer, but we will be healthier doing it.

Another often ignored factor is the effect of the medications we take to deal with the diseases and ailments we suffer as we get older. There is evidence that at least some of these medications may cause or contribute to the forgetfulness, confusion, and disorientation some people experience as they age. The suspect medications include tranquilizers, blood pressure treatments, and even aspirin. The older we get, the greater and the less predictable the effects of these medications. And it is often combinations of medications that create the problems. If you use any of these medicines, be sure to ask your doctor how they might affect each other—and you—in combinations.

On the other hand, some medications alleviate aspects of the mental deterioration that can be associated with getting older. Nimodipine, a calcium-blocker used to improve cerebral blood flow in older people with cerebrovascular problems, has been found to offset declining learning ability. Phosphatidyl-serine may be a chemical that can regenerate failing memory banks.

What all this will mean is that we will have to rethink what getting older means. In some businesses—advertising, for example, or science—getting older has traditionally meant becoming less able to be creative, inventive, and energetic. Tell that to Leonardo da Vinci. That image of mental as well as physical decline over forty-five or fifty or fifty-five or even sixty will no longer hold water as more and more people stay vigorous, vital, and sharp into their sixties and beyond.

So, as others around you in the office show signs of aging, be patient and help them. Their difficulties will eventually become your own. And until they do, exercise, eat well, and take real good care of yourself, because the twenty-first century is right around the corner. And by the time you get to the other side, we may have a cornucopia of cures for what ails your aging bones. Then you will be well prepared to begin your third, fourth, and even fifth careers.

5 *You Are What You Eat: Wellness, Nutrition, and Diet*

"Gourmets," someone once said, "live to eat, but the rest of us eat to live." What we are also coming to discover is that *what* we eat can determine *how* we live—and how we work.

Every day now, newspapers and magazines publish more articles on the relationship between diet and health. It's hard to believe that after twenty years of this, most people are ignorant of the relationship. But although there have been some very significant changes in the American diet over those years—for example, the shifts away from red meat, eggs, and high-fat dairy products—Americans still ingest an overabundance of fat, and far too many Americans are obese.

A study done by the Nutri/System Company in 1989 revealed that under two-thirds of Americans scored satisfactorily on a test of nutritional knowledge. Among the findings: over 40 percent thought catsup was a vegetable.

One reason for the growing focus on the link between health and eating is the aging of the Baby Boom. That dominating segment of the workforce is finding, as they get older, that exercise may keep them youthful, but it won't keep them young. Inexorably, no matter how determinedly you exercise, you see the creeping advance of age: the waistline thickens, the hairline recedes, the eyes dim, the cellulite convenes, and the wrinkles stay on. In addition, demands of family and job make it harder to maintain rigorous exercise schedules. So, as younger and more fit workers enter the office and workplace, Boomers are reminded of their flagging physiques.

Workers looking for the fount of youth are becoming as serious about nutrition as they have been about exercise. And they are seeking not only youthfulness but wellness. Americans in particular are obsessive about health; good health is one of the qualities of life we seem to feel we are entitled to. And as scientists become more able to relate nutritional intake to overall health and productivity, this focus is reinforced.

A Sampling of Science

- A study reported in the prominent British medical journal *Lancet* showed that children's performance in school could be improved measurably by regular use of vitamin-mineral supplements. The students showed marked improvement in nonverbal intelligence, some of which seems to be related to physical growth. (Verbal intelligence, on the other hand, may be more closely tied to cultural and educational factors.) It should be noted that the supplements had this effect on children whose diets *seemed* nutritionally adequate. According to two British researchers, David Benton and Gwilym Roberts, the diets of such schoolchildren, in reality, tend to be deficient in vitamins, particularly thiamine, folic acid, and vitamin D, and in minerals.
- Recent studies have also linked nutritional deficiencies to behavior. Stephen Schoenthaler of California State University did a test on juvenile inmates at cor-

rections institutions. He took away their high-sugar-content junk food snacks and sodas and replaced them with fruit juices and low-fat high-nutrition snack foods. Antisocial behavior—aggressiveness, fighting, etc.—was cut in half.

• A study at the National Institute of Mental Health found that aggressive behavior in animals was linked to low levels of the chemical 5-hydroxyindoleacetic acid (5-HIAA). Subsequent research showed that aggressive humans also had low levels of that same chemical. A diet with increased vitamins and minerals and reduced sucrose and refined carbohydrates can raise levels of 5-HIAA and lessen aggressiveness. The next time your boss or co-worker gets testy, chart his or her sugar consumption. Maybe offering a celery stick is a better solution than getting upset.

• A study the DuPont company did of its workers showed a significant difference in coronary heart disease rates between white collar and blue collar workers. From 1957 to 1983, CHD rates for white collar workers declined by 37.6 percent; for blue collar workers the decline was only 18.2 percent—less than half. Differences in diet and life-style accounted for this startling variation, according to scientists who studied the results, and could clearly have a major impact on the cost to you and your company for health insurance premiums.

• Most medical authorities believe there is a link between CHD and high levels of LDL cholesterol in the blood. (HDL cholesterol, on the other hand, is considered beneficial.) Recent studies seem to substantiate claims that oat bran can have a moderately beneficial effect. Adults who ate a lot of oat bran every day saw, on the average, a 2 to 3 percent reduction in cholesterol levels. The higher the individual's cholesterol level, the greater the reductions: People with cholesterol levels of over 229 milligrams per deciliter of blood experienced drops of up to 7 percent. However, the oat bran muffins that appear in the company cafeteria or on the coffee cart don't seem to work. Apparently, egg yolks and other fats in muffins offset the effects of the bran itself. So don't think that because you've selected the oat bran over the blueberry that you've done the right thing. The fact is, no one factor will make you healthier. Oat bran works best if it is part of a regimen that includes low-fat, low-sugar, low-sodium foods as well as exercise.

• As noted in chapter 4, there is growing evidence that diet can affect the symptoms of menopause. Japanese women, apparently because of the soy products they eat, are less bothered by such symptoms as hot flashes than are American women. As nutrition authority Jane Brody explains, diet can help women compensate for some of the physiological changes that come after menopause. A diet that contains lots of fruits, vegetables, and whole grains can help lower cholesterol. Soluble fibers, found in such foods as oats, barley, carrots, grapes, and apples, are especially helpful. Foods rich in calcium—low-fat dairy products, for example, and dark-green leafy vegetables—can help offset bone deterioration. Too much protein can counter this, so older women should not eat too much meat. Overindulgence in alcohol can also undermine the effects of calcium-building foods.

• Dr. Joseph D. Beasley, who did a landmark report on health for the Kellogg Foundation in 1989, says that "we are rendering ourselves toxic and malnourished" and that much of our health problems can be attributed to years of inadequate nutrition. Beasley believes that malnutrition is increasing in the United States because modern mass-produced food is both more contaminated with toxic chemicals and less nutritious.

His contentions seem to be borne out by the results of

a fifteen-year study done for the United Nations Environmental Program. This study, "Global Assessment of Soil Degradation," reports that much of the world's agricultural soil has lost substantial amounts of natural fertility and, therefore, requires increasing amounts of chemical fertilizers.

Although U.S. soil remains relatively fertile compared with some other parts of the world—e.g., Europe and Asia—remember that our food supply here is global. The result of poorer soil is food weakened in nutritional content.

What's a Person to Do?

Given all this, what can you do? (Keep in mind that being overly obsessive about food, like being overly obsessive about health, could ultimately be unhealthy. Reason and moderation, as the ancient Greeks kept saying, can guide you to a good regimen.)

Many people are turning to organic foods, those grown or produced with minimal or no help from man-made chemicals. Health food stores and restaurants that feature such foods are cropping up everywhere. There are even organic snacks now—blue corn tortilla chips, for example, and organic popcorn. Chances are, you've seen trail mix and fresh fruit displayed for some time now in the company cafeteria.

People are also becoming more aware that there are individual relationships with food: some of us are allergic to some foods, and our allergic reactions can vary greatly. Milk and milk products don't agree with some of us. There are also people who are allergic to wheat. In both cases, substitutes are now widely available. If we are lactose-intolerant, for example, we can use yogurt; we can also get pizza made with soy instead of cheese. Other grains can be substituted for wheat in bread: spelt, for example. Bakeries are coming out with bread, and even bagels, baked from spelt. There are wheat-free pastas, as well.

Other foods that can cause allergic reactions include eggs, fish, legumes, some nuts, and seafood. All of these are rich in protein, and it is protein that causes allergies in people. Indeed, one of the concerns about genetically altered foods is that transferring a protein from one food to another may make that second food allergenic. In many foods that cause allergic reactions, the specific protein responsible has not yet been identified.

There's also renewed interest in food supplements, particularly vitamins and minerals. The evidence linking these supplements and improvements in health is more impressive all the time. Beta carotene, for example, seems to play a significant role in reducing coronary heart disease and lung cancer.

Supposedly, if you eat a balanced and nutritious diet, you get all the vitamins and minerals you need. But few of us are so consistently good about our diets. And, as pointed out above, even balanced diets may have nutritional deficiencies. Supplements, however, cannot totally substitute for a good diet. There are elements in food that are not too easy to compensate for, such as fiber. Other useful and even necessary elements in food that we are now learning about include:

- *Sulforophane*—found in Brussels sprouts, cauliflower, carrots and other foods, it may be a tumor-blocker.
- *3-n-butyl phthalide*—found in celery, it can lower blood pressure and reduce cholesterol.
- *Resveratrol*—found in red grapes and red wine; may also reduce cholesterol.

Another important dietary point is the relationship to aging. As we get older, we need fewer calories (particularly because our work tends to become even more sedentary), but we still need all the nutrients. Indeed, we need even more of some nutrients. Postmenopausal women, for example, should have 1,500 milligrams of calcium, while the recommended daily allowance (RDA) for younger women is no more than 1,200 milligrams.

There are some who try to do more with supplements than get themselves good nutrition. These people use vitamins and minerals, combined with certain medications, as drugs—to improve their intelligence, increase their stamina, enhance their feelings of well-being, etc. This use of so-called smart drugs is particularly popular with aging Baby Boomers who believe they can improve themselves and remain young and who are increasingly disappointed in the results of exercise and diet.

The medical community and the Food and Drug Administration are concerned that the claims made for many of these smart drugs are unwarranted—and that there may be unhealthy side effects from continued use. (Apparently, many of the products contain caffeine and/or sugar, which can account for the rush of energy some people experience when using them. Accordingly, one name for some of the products, says *The New York Times*, is "high-tech coffee.")

While the scientific community has its doubts about smart drugs, there is an increasing amount of research into using food to heal or prevent diseases.

Government and business laboratories are experimenting with "designer foods"—identifying and extracting from foods the chemicals (called phylochemicals) that have disease-fighting capabilities and then adding or reinforcing these chemicals to make foods more powerful disease fighters.

For example, scientists at the University of Pittsburgh have identified a chemical, pyrroloquinoline quinone (PQQ), in dairy products that can prevent brain damage from strokes.

This subject gets more complicated—and confusing—everyday. What's right or wrong for you? What will help you? What do you need to live and what do you need to be healthy and what do you need to be happy and productive at work?

Clearly, the answers to these and other questions concerning nutrition and health depend at least to some extent upon your individual physiology. It is important to recognize that there are limits to what you can do—people who eat right and exercise might still die prematurely of heart disease. High blood pressure, for example, seems to run in families, indicating a genetic base. If you are susceptible to it, proper diet and exercise can only help you control it.

It is also important to know what you are doing: don't rush into fad diets or unfamiliar medications; they can be dangerous.

The first thing is to know as much as you can about the key nutrients and why you need them. This information is available from many sources. The U.S. Department of Agriculture Human Nutrition Information Service has a publication, *Nutritive Value of Foods* (Home and Garden Bulletin #2). Food companies can also provide useful publications. One example is the Consumer Center of the General Foods Company, 250 North Street, White Plains, New York 10625.

In the Mood for Food

Sarah Leibowitz, a professor of neuropharmacology at Rockefeller University, says that the key to what we eat and what it does for and to us is brain chemistry. According to Dr. Leibowitz, the brain's purpose is to achieve energy balance, and it goes about doing this by regulating appetite and metabolism through juggling chemicals in the brain. All of this shapes our moods, our energy, and our levels of stress. New research shows, for example, that the craving for carbohydrates starts off strong in the morning and wanes during the day; the opposite is true for fats.

One part of the hypothalamus, the paraventricular nucleus (PVN), appears to be the part of the brain that acts as the thermostat for appetite. It regulates production of the chemicals that stimulate or repress appetite. In the morning, when blood sugar levels are low and energy depleted, the body needs a quick fix of glucose. Carbohydrates convert quickest to glucose, so that's what the PVN asks for. Late in the day, the body needs to store up energy for the night, so the PVN asks for fat, the means whereby calories can be stored for future use.

When you eat certain foods may be important for other reasons, too. If you have a big meeting at 10:00 A.M., don't eat a sweet roll at 9:00 A.M.; the pick-me-up from the sugar could turn into a slow-me-down an hour later. That could dampen your alertness, negotiating skills, and presentation style. For that same meeting, don't drink too much caffeine; you could become hyper with some possible negative effects on your judgment ability. Many people diet by skipping breakfast; obviously, this is counterproductive. The body needs an infusion of energy in the morning after the long night. Lack of that energy will be reflected in declined mental and physical acuity.

As the links between diet and health and diet and job performance become clearer, employers as well as employees are acting to produce both better health and greater productivity. Some companies go so far as to ask their employees to limit fast food consumption. Rising health care costs are spurring many companies to urge or even force employees to eat better—and some organizations even give financial incentives for employees who reduce cholesterol or lose weight.

The human digestive system, as rugged as it is, is also very sensitive. Things other than food can affect it; as the great Colombian author Gabriel García Márquez wrote, "Love . . . is a disturbance of the digestive system." Our emotions, and our emotional responses to events, can lead us to overeat or to eat (or drink) the wrong things. Traumas—being fired, for example, or being outraged by the boss—can offset the beneficial effects of good food.

A good diet and physical fitness are both very important, but what most health gurus fail to mention is something simple and inexpensive: the art of listening. Jim Perry of our company has studied the relationship between diet and health for years and has made himself into an expert on the subject. (He's awfully healthy and fit.) He says that the long journey to physical well-being begins with listening to your body, and if you listen closely enough, it will reveal many interesting secrets. It will tell you what it needs and when it needs it. The trouble is that our eating patterns tend to be shaped by other forces, and our habits prevent us from hearing our bodies.

For example, most of us have a pattern of three sizable meals a day, each at a regular time, sometimes supplemented by between-meal snacks. This pattern is often, if not usually, accompanied by feelings of hunger, tiredness, etc., at intervals during the day. But we often ignore these signals, even though they can affect our work, because eating at those odd times is not what we have been trained to do. The fact is, the eating pattern known as "grazing"—frequent small meals or snacks during the day—may be more in tune with what our bodies need. And this could help make us more consistently energetic and alert at work.

Doing the Research

One of the ways to take control of your own life and health is to get the information you need about diet and nutrition for yourself. Be warned. That is not as easy as it sounds. Here's what happened when we tried to get a particular bit of information.

A friend who is a food authority for a well-known publication suggested that we ask the Center for Science in the Public Interest and *Prevention* magazine. Neither organization had what we wanted. *Prevention* said we could get it from the Food and Nutrition Information Center of the U.S. Department of

Agriculture and graciously provided the center's telephone number.

When we dialed that number, we got a recorded message saying the number was no longer in service. No alternative number was given. We then called the Department of Agriculture in New York City and were given another number for the Food and Nutrition Information Center. That number, too, was out of service.

We then called the Agriculture Department in Washington and got another number. That number worked. We got through to a pleasant woman who said that FNIC did not have the nutritional information we were after; it administered the Food Stamp program. She referred us instead to the Agriculture Department's Human Nutrition Information Service. She provided the telephone number for HNIS and helpfully offered to transfer us. When she attempted to do so, we were cut off.

But we had the number, so we called HNIS. The receptionist, another pleasant woman, told us whom we wanted to speak to and said she would transfer us. Older and wiser by then, we requested the telephone number of that person. And that was smart, because we were cut off again.

The person we finally got through to at HNIS told us there was indeed some material of the kind we wanted available, and it could be ordered from the Government Printing Office. She gave a New York number for the GPO, which we immediately called. It was busy.

After several tries, we got through and we ordered the material. Unfortunately, when it arrived, it wasn't quite what we had wanted.

6 *Get Smart: How We Think and Learn*

In 1988, *The Wall Street Journal* reported on an interesting attempt to copy one man's unique knowledge. The Southern Edison Company was about to lose one of its key engineers, Thomas Kelly. Kelly was the company's expert on a troublesome dam in the Sierra Nevadas; he knew all about that dam and all its complications. So the company spent two years and more than $300,000 to create a computer program—an "expert system"—that would contain all of Kelly's knowledge and understanding of that damn dam.

The project was dropped, and Thomas Kelly went on working for Southern California Edison.

A major difficulty was that Kelly didn't necessarily understand what he knew, or how he knew it, nor was he able to repeat it in the logical, step-by-step way necessary to incorporate it into a computer program.

Human intelligence is hard to define and harder to pin down. It includes instinct, intuition, insight, and the ability to leap over gaps in information and somehow come to understanding.

The truth is, we know less about the mind than we do about the universe. For too long, we relied on explanations of the mind and its workings that are no longer seen as scientifically valid. The conventional wisdom about the mind until recently was equal to scientific thought about the universe before Galileo. But now scientists are taking another look at the assumptions behind the thinking—at the theories of Descartes and Freud, for example—and are approaching them with more skepticism.

One such scientist is Gerald M. Edelman, who has written a controversial book, *Bright Air, Brilliant Fire,* that sheds new light on the mind and the brain. Edelman is director of the Neurosciences Institute and chairman of the Department of Neurobiology at the Scripps Research Institute. He won the Nobel Prize in 1972.

According to Edelman, we are now just beginning to see a revolution in neuroscience, one that will ultimately reveal to us how the mind really works—and that its basis is biological. Mental processes begin with biology; how the brain is organized biologically—its morphology—determines how what we call the mind works. So, as Edelman says, we can't begin to describe the mind accurately without a "detailed biological description of the brain."

What Is the Brain?

The brain has five basic parts—the cortex and its four appendages: the cerebellum, the hippocampus, the brain stem, and the basal ganglia. These are connected to each other by neurons and to the outside—i.e., the sensory organs, glands, and muscles—by other neurons. These neurons communicate with each other through synapses, which are, in effect, electrical *connectors.* Each human brain is made up of approximately 10^{11} (100,000,000,000) cells and about 10^{15} (1,000,000,000,000,000) connections. This basic structure is similar for all human brains, but each individual brain's "map"—how all the neurons are connected—is different from all other brain maps.

The brain is arranged into two basic nervous systems. One involves such aspects of behavior as appetite, sexual drive, and self-defense. This is the network for operating the heart and respiratory system, digestive system, and various basic body cycles. The second, known as the thalamocortical system, receives signals from outside and controls muscular response to those signals. The two systems are linked so as to coordinate behavior to external circumstances. This, significantly, forms the basis for learning.

The higher brain functions are those connected with what we call intelligence—speech, for example. There are three such broad higher functions: categorization, memory, and learning. And it is these that, along with the development of the vocal tract, led to the unique human capability for language.

What are known as diseases of the mind—those conditions that interfere with, warp, or diminish normal operation of higher functions—are, according to Edelman, physiologically based. This hypothesis can explain why drugs that change how synapses work can effectively correct mental disorders. Schizophrenia, characterized by such symptoms as hallucinations and delusions, is an example of a disorder that can be brought under control by medications that appear to mend or realign the brain map. Even phobias (fears) can be treated successfully by antidepressant drugs, such as phenelzine.

As Judith Hooper and Dick Teresi point out in their book, *The 3-Pound Universe*, what we have always called "mental" problems—including delusions, obsessions, and even strange beliefs—are not just mental; they are related to brain morphology.

Obsessive compulsive disorder (OCD), which affects perhaps as many as one in forty people in the United States, is a condition that makes its victims engage in bizarre behavior. An OCD victim, for example, may wash his hands over and over again, dozens of times a day, out of an overwhelming fear of contamination. Research shows that OCD, like other mental aberrations, results from a miswired brain circuit.

Behavioral scientists at UCLA, led by psychiatrist Lewis Baxter, have used behavioral therapy to treat OCD sufferers. What is intriguing about Dr. Baxter's work is that the therapy appears to produce metabolic changes in the brain. That is, the brain, perhaps as a consequence of the therapy, appears to rewire itself, a response similar to that achieved by drugs.

Intelligence

Society has generally seen intelligence, or high IQ, as the opposite of mental disorders. As we learn more about the brain, however, it is becoming apparent that this traditional belief is not necessarily so. Dr. Aubrey Milunsky, a noted geneticist and pediatrician, points out that learning disorders, such as dyslexia and even autism, may be linked to superior intelligence. In the movie *Rain Man*, the autistic character played by Dustin Hoffman had mathematical abilities approaching genius.

Dyslexia, autism, and other conditions are manifestations of brain disorders: some part of the brain has its wiring crossed or wrongly connected. Some of these conditions are caused by damage to the brain, perhaps during birth, that reduces oxygen flow. But there are also strong indications that such conditions can be genetic in nature.

About 15 percent of the U.S. population (a figure that also applies to most Western countries) is affected by some form of dyslexia—primarily difficulty in reading. Such people used to be regarded as low in intelligence or even retarded, but IQ tests show that is not the case. What dyslexics have is, basically, an electrical malfunction in the brain circuits. And it appears to be hereditary.

It also affects males more than females, reinforcing our understanding that male and female brains are different (see chapter 3).

Scientists are only beginning to identify specific genes associated with higher learning functions. Recently, two teams of scientists, one at MIT and the other at the Howard Hughes Medical Institutes, removed one gene from mice and adversely affected the ability of those mice to learn and remember. These experiments provide the most convincing evidence so far that the functions of certain genes are the basis for higher learning functions. With the gene, one that codes for a particular enzyme, removed, the mice were unable to use visual clues to escape from a maze.

Studies such as these are bringing the relatively new science of molecular genetics to the study of the brain. Even leading psychologists, such as Tom Bouchard, of the University of Minnesota, increasingly see genes as the basis for human behavior. Bouchard, widely known for his studies of identical twins, also believes levels of sex steroids in the womb (see chapter 3) have a significant effect on later behavior. Twins in the same womb may have variations in hormonal levels.

Behavioralists and geneticists increasingly say that specific genes cause or affect such brain functions as intelligence and personality and such conditions as schizophrenia. They expect to identify all these genes before long and to learn much more about how they work. They also expect that genetic engineering will enable us to correct genetic flaws in intelligence and behavior. Nick Martin, an Australian geneticist, predicts that we will soon be able to design classes in school for children based upon their individual genetic makeup.

In another area of research, scientists are learning more about how different parts of the brain are involved in different aspects of intelligence. Dr. Justine Sergent, a neurologist at

Canada's McGill University, found that the part of the brain responsible for certain musical skills—sight-reading and playing an instrument—is separate from (but parallel to) the part responsible for verbal skills. So Maurice Ravel, a great musician/composer, was able to keep his musical skills for some time after a brain disease destroyed his ability to speak and write.

A logical question, in light of this, is, is intelligence inherited? The answer to that one, unfortunately, is not so simple. Earlier in this century, the consensus answer was yes. Over the years, however, this conclusion was challenged—not only by scientists but by idealists and politicians who felt that such a conclusion was inconsistent with our ideals of equality, opportunity, and progress.

Recent research on identical twins by Bouchard and others suggests strongly that intelligence is largely if not entirely genetic. But most such studies are criticized by some scientists who claim they undervalue the role of environment. On the other hand, studies of adopted children tend to show more impact of environment on intelligence.

One difficulty is with how we measure intelligence. The primary method is the well-known IQ test. But there is growing questioning of whether the standard IQ test accurately measures intelligence; indeed, many now even question whether there is an "intelligence" that is measurable by one test. IQ tests may be reasonably good at measuring verbal and numerical skills and inductive reasoning abilities (although many challenge even this), but there are other forms and manifestations of intelligence that they don't measure. In fact, the widely heard (and cynical) comment about IQ tests is that they accurately measure the ability to do well on IQ tests. Somewhat less cynical, but also widely held in business, is the belief that IQ tests are good indicators of how young people will do in school—and nothing more.

The tests are also faulted for what the critics call cultural, racial, and socioeconomic biases.

Professor Howard Gardner of Harvard postulates that there are seven kinds of intelligence, including such things as athletic ability, musical ability, and interpersonal skills such as empathy. Sales managers know that there is such a thing as sales intelligence—persuasiveness and knowing when to close the sale. And in sales, persistence and determination are great assets, but there is no way to test for them. These, too, are a form of intelligence. It's difficult to teach these things, but good salespeople just seem to know them.

Gardner's theories are beginning to show up in some school curricula around the country. These schools try to give as much emphasis to the other forms of intelligence as is given to verbal and mathematics skills. Lauren Resnick, of the University of Pittsburgh's Learning Research and Development Center, says that schools have to focus on helping children develop the skills they will need in later life. Dr. Resnick labels these "practical intelligence," as compared with the "school intelligence" they encounter in their schooling. School intelligence, according to Resnick, consists mostly of "pure thought" activities. Practical intelligence, on the other hand, entails knowing how to use available tools, how to work with other people, and how to make the abstract specific.

In other words, IQ isn't everything. Seymour Epstein, a psychologist and professor at the University of Massachusetts, goes so far as to say flatly that "IQ and success in living have little to do with each other." Epstein says that "emotional intelligence" is much more important. The individual who knows all the facts but freezes when he appears on "Jeopardy" lacks emotional intelligence. He can't manage his emotions, so his fund of information becomes useless.

Most of us, maybe even all of us, have had the experience of doing the wrong thing—the "dumb" thing—in a pressure situation. Trying to make a good impression on a member of the opposite sex, for example. Someone who does that all the time instead of just once in a while lacks, as Epstein puts it, the ability to respond effectively to life.

Epstein tries to measure what he calls "constructive thinking"—how people respond to challenging situations when they need to control their emotions. Everyday life, in other words.

The ability to choose the appropriate response to circumstances or events is an important skill. Whether to flee or fight; whether to laugh or cry. Good management is often good judgment, being able to make the right decision under pressure. It happens when the brain, having analyzed the information it has, gives a clear, sharp signal. When the brain, perhaps overwhelmed by the quantity of information or the possibilities of response, cannot effectively choose, then the individual doesn't know what to do.

An example: Two co-workers at the company picnic are chatting by the edge of the crowded swimming pool. Suddenly, they both see a child struggling in the water. The first person reviews the options—should I get the child?; should I alert the lifeguard?; should I shout something?; if I shout, what should it be? The second person immediately jumps in and grabs the child before the first is through reviewing the options. What makes the difference? And does the way they react here mimic the way they make decisions back at the office?

Is the Mind a Computer?

One of the great fallacies of current thinking about the mind, as Edelman and others have pointed out, is to view it as a kind of computer. The eminent English mathematician Roger Penrose says that comparing the mind to a computer is absurd. As he points out in his book, *The Emperor's New Mind*, what goes on in the mind is not "computable." In agreement, Edelman says that the way the mind works does not even come close to resembling computer-style logic and computation. For one thing, brain memory, unlike computer memory, is not precise. A computer will recall from memory exactly what was put into it; humans filter memory through perceptions, prejudices, and a host of other factors.

The basic difference is that the humans are active participants in the learning process. We don't just read input, the way a computer does; we interpret it, applying what scientists call "cognitive mediation." A machine can perform certain functions of logic, but only if it has been specifically programmed to do so.

Perhaps an even more striking difference is that, according to some new findings, the human brain may not be "hardwired." It was, until recently, widely believed in scientific circles that most of the brain's information circuits are permanently fixed, just like a computer's. But more and more evidence indicates that the brain can constantly rewire itself. Recent work by such scientists as Michael Merzenich of the University of California and Jon Kaas of Vanderbilt University shows that traumas, such as the loss of a limb, actually result in compensatory rewiring of the brain. The brain appears capable of making new maps as needed. This adaptability of the brain may well be a key to understanding how we learn.

We translate information into images. To a computer, the numbers in its memory are reality; to a human, the words in his or her memory are only symbols that create images. The word "chair" is not the reality of a chair; it is the symbol of a chair, and the word enables us to picture the actual object.

Learning

That's why some of the most effective learning devices for us are metaphors and similes. When we say of a woman that "her hair is like silk," we create a mental image both striking and concrete. We can both see and remember strongly. The best teachers and the best writers enable us to see vivid images and therefore to learn. The enduring popularity of Shakespeare owes much to his extraordinary genius in creating memorable images:

> "Now is the winter of our discontent made glorious summer by this sun of York."
> "Shall I compare thee to a summer's day?"
> "The quality of mercy is not strained; it droppeth as the gentle rain from Heaven . . ."
> "Night's candles are burned out, and jocund day stands tiptoe on the misty mountain tops."

Also unlike a computer, information stored in the brain tends to decay over time. The rate of decay varies by individuals. Some people, those we label as having photographic memories, can recall with astonishing accuracy over long periods of time. Most of us, though, can't do that. With us, the rate of decay depends on how we learned the information, how important we think it is, and how often we use it. Repetition in learning, as done in school, retards decay. That's why we can often remember accurately bits of poetry we learned in grammar school, but hardly anything of a conversation we had yesterday.

Neuropsychologist Robert Desimone, of the National Institute of Mental Health, uses the phrase "the physiology of memory" to characterize the new way of looking at how we remember. According to Desimone, the brain undergoes actual biological changes as it learns, and these changes determine memory. For example, neurons will behave differently depending upon how familiar or unfamiliar specific visual stimuli may be. The more familiar the stimulus, the lower the activity level as the neurons rely more on previously stored information.

We are only now beginning to learn about how to write

instructions that match the way the reader's brain learns. Studies suggest that when a printed manual contains an overview, many readers would work from that rather than make the effort to read the rest of the instructions. And people want a "cookbook" format for their instructions—broken down into quick bits that they can visualize and manage. Too much information is often felt to be a burden on their memory or understanding, thus hurting rather than helping their chances of learning. And if people have to read instructions on a screen, the "keep it simple" rule is even more important; otherwise, without page numbers and the ability to leaf through pages, they feel they are getting lost in the text.

Other key factors in learning are mood and motivation (see chapter 13). We learn more and better when we are highly motivated. If we're less motivated, we tend to give up quicker, and our lack of interest in the subject makes us reluctant to concentrate enough.

Depressed people have considerable difficulty in learning. In one recent experiment, depressed people were 30 percent less efficient in learning. Anxiety also has adverse effects. We all know of smart people who don't test well because of "test anxiety."

Recent studies have shown that relaxation methods can help offset these negative impacts on learning. Meditation and biofeedback, for example, can alter mood in positive ways and improve both learning and recall. Chronic high stress levels can greatly hurt learning ability; indeed, there are reasons to believe that this can actually destroy brain cells (see chapter 10).

One of the most intriguing concepts to come to light recently is that the brain picks up sensory stimuli from the environment in which learning occurs—*and these are permanently associated with the information.* So, for example, if you learned algebra from a teacher who always wore a strong lilac scent, the scent of lilacs thereafter will improve your recall of algebra. One theory is that you should study in the room where you will be tested, on the grounds that the familiar sights and odors will stimulate memory.

But it is important to remember that not everyone learns the same way, or equally well. We Americans tend to have the optimistic belief that if we want, for example, more engineers, we only need to send more people to engineering school. We may get *more* engineers, but that doesn't necessarily mean we'll get *capable* engineers. (In fact, the lowered level of standards may lower the quality of *all* engineering as a result.) What we are learning from all these new scientific approaches to the brain is that learning requires more than just the availability of information; it requires a brain with the specific higher functions that can best use that information.

It also means that where skills needs are not being met by the skills pool, organizations have to be more creative in searching out alternatives. For example, according to recent surveys, maybe 10 percent or more of American workers are "functionally illiterate." They have difficulty reading and writing. This doesn't necessarily mean, however, that they are unintelligent or that they lack skills. Functional illiterates may only lack certain language and/or reasoning skills, making it difficult for them to absorb information from texts. In any number of instances, creative employers can help them overcome their handicaps. People who can't write, for example, can deliver oral or taped reports. People who can't read can learn to use computers that train operators with symbols rather than words. Creating teams where some individuals compensate for skills lacked by others can also work.

Some years ago, we had a high-tech company client that used COBOL, the most advanced computer language at that time. The company recruited the top graduates trained in COBOL from some of the most prestigious technical universities in the United States. But within a few years, COBOL was replaced by more advanced programming languages. So now the company wound up with a large pool of people who were being paid high salaries to "keep the computer files in order"—high-tech janitors, you might say. Not one was able to shift from the thought pattern that was used in COBOL to the thought process needed for the next-generation language. Perhaps the company would have been far better off if they had done their recruiting not on the basis of a highly specialized skill, but a broader set of competencies. For example, if they had screened graduates from the Juilliard School of Music, they might have found potential employees who were very talented with computers (music is often based on computer technology today), who could get all sorts of computers to talk to each other, and who were capable of thinking abstractly in many languages and symbols, not just one pattern. In the rapidly changing world we live in, it may be the *competency* to learn new skills rather than the skills themselves that we need to focus on.

American business spends more than $30 billion a year in alternative learning efforts. These efforts range from sleep learning—e.g., playing tapes to people while they are sleeping—to various forms of creativity training. All this, in spite of the fact that the National Research Council, after a three-year study, concluded that most such efforts are not scientifically valid.

In recent years especially, New Age or human potential programs have been warmly received by many businesses. They have sent people out into the wilderness to learn how survival skills have relevance at work (perhaps validating the often heard statement about the workplace: "It's a jungle out there"). They have probed the psyches of employees in a variety of ways* and subjected them to intense emotional

*Heightened interest in the brain—and particularly in improving it—has attracted entrepreneurs. A recent issue of *Psychology Today* magazine offered a self-improvement tape that promised a "supercharged brain" in only twenty-eight minutes.

experiences. There is very little, if any, understanding of what the return is to companies on their investment in these activities.

It is not just business jumping on the bandwagon and exploring every learning fad that comes along. Schools, too, are trying all kinds of things. The noted educator and critic Jacques Barzun castigated the educational system for all its faddism. Thinking, says Barzun, cannot be taught; it can be drawn out by giving students material they have to think about. Couldn't we take this to mean that perhaps the best way of getting good thinking in the office is by giving ourselves the right stuff to think about?

Brain Food

Exercising the brain, as we are now discovering, requires an understanding of the biology of the brain. Assumptions, conventional wisdom, and unproved theories need to be taken with a grain of salt unless and until they can demonstrate that they are consistent with the growing body of scientific data about how and why we think and learn.

One very promising avenue for further study is the relationship between the brain and food. Scientists are identifying what they call "key nutrients" that can beef up brain performance. These include all the B vitamins, vitamin C, the amino acids tyrosine and tryptophan, the minerals iron, copper, and zinc, and the electrolytes calcium, magnesium, potassium, and sodium. A normal balanced diet includes all these, but a complete vitamin-mineral supplement can help if you're not sure you are getting all the nutrition your brain needs.

A growing number of people believe that brain functions can be improved—and declines in brain functions offset or reversed—by the use of "nootropics": drugs and chemicals that interact with the biology of the brain. There are, for example, the so-called "smart drugs"—usually combinations of vitamins (particularly the B complex, C, and D), amino acids, fructose, choline, and stimulants such as caffeine. It is believed that such substances can improve such brain functions as concentration, memory, and problem solving—and that they can delay some of the effects of aging on the brain.

Pharmaceuticals prescribed for other purposes are also used as nootropics. These include Dilantin, widely used to treat epilepsy and brain seizures. Dilantin appears to act by stabilizing electrical activity in the brain. It is credited by its advocates with increasing IQ and improving both memory and verbal intelligence. Supposedly, it also eases obsessiveness and even counteracts jet lag.

Some natural substances are also used as nootropics. Ginseng, for one, which is believed to improve memory and other brain functions. All these substances are believed to work by enhancing the neurotransmitters—the connectors in the brain. It is claimed that smart drugs, such as Piracetem, can even restore neurotransmitters that have been damaged by use of cocaine and other illicit drugs.

Many scientists scoff at all this; they say none of it has been scientifically established. And the FDA is opposed to the use of these substances for mental enhancement for otherwise healthy people. But other scientists are less negative. They feel that nootropics probably do no harm and may even do a little good.

At this point, moderate use of nootropics may be like what your mother used to say about chicken soup: it can't hurt. And as the office of the future becomes a more competitive, highly charged place, brain food may be just the ticket to keeping your sanity *and* your competitive edge.

7 Ergonomics: We Can't All Be Five-Foot-Ten-Inch Males

It's the end of another workday, and you hurt. Your arms, wrists, and neck feel stiff, your knees are tight and your head is throbbing. Your's and millions of other people's!

> Spring, 1991. Several women were having lunch together, and the conversation went as follows: J., a stockbroker, complained that her legs were not what they used to be, especially her left leg. She was only forty-five, and loved to walk and hike and ski, but that was getting harder and harder. She went to two specialists and for a series of tests, but they showed nothing and offered no relief. Was it aging? B., a branch bank officer, confessed she had a similar problem, and she was only thirty-four. She had been considering doing physical therapy. Her doctor said the problem seemed to originate from her knee. E., a copy editor, said it wasn't her legs that hurt so much as her neck and lower back. For the past two years, she had been experiencing on and off chronic pain. She had been to a chiropractor, joined a health club to swim, bought a very hard mattress, and occasionally had to take strong prescription painkillers. T., an executive assistant, said

> she had a good chiropractor to recommend, because she had the same problem. "Could this be early arthritis?" asked B. "Isn't that an inherited problem?" asked T. With that, the discussion turned to mothers, and none could recall her mother having such chronic and severe problems this early in life. "Is it just a woman's thing?" asked T. E. laughed. "You should see the guys lined up in the chiropractor's office!"

If you often leave the office with stress, backaches, headaches, or other pains, consider that these kinds of symptoms may be your body's way of saying it can't adapt to the physical aspects of your workplace. The problem may be that your workspace and office equipment simply were not designed for you. And that's because most workplaces are designed for the average white male of about five feet ten inches.

But height is just one of many physical problems that may be making it harder for you to do your best job, or to keep it at all. While sociological studies have indicated as much as double the rate of turnover and failure among minorities and women in the mostly white male cultures of many U.S. corporations, we need to know much more about how physical differences among groups can hinder people in the increasingly diverse workplace.

Taking the Office to the Fitting Room

There is a fairly recent but growing body of research that studies people at work: ergonomics.*

The U.S. military started the field of ergonomics during

*One source defines it as "the application of physiology, psychology and industrial engineering to improve the work system including the person, job, tools and equipment, workplace and work space, and the immediate environment."

Another definition states that ergonomics uses various sciences to study "human needs, characteristics, abilities, and skills as applied to the design, production, and management of products or services, within a given work environment and social surroundings."

World War II by applying experimental psychology to solve various problems between people and their machinery that kept arising in combat (and, it later found, in industry as well).

Why do we need ergonomics? Because, let's face it, the engineers who designed your computer, telephone, fax, copier, desk, chair, lighting, heating and cooling system, windows, office layout, conference rooms, and just about every other piece of equipment and element in your workplace were generally more interested in the technical specifications and capabilities of the hardware than in your comfort, familiarity, and ease of use.

Most of us work in a human-machine system. And while few of us like to be compared with machines, we are just one part of the system including hardware and software, personnel, operating procedure, and technical data.

Applied properly, ergonomics can improve our safety and health, reduce costs and increase productivity, and make work more comfortable—goals that should be shared by workers, employees, and insurance companies alike.

Nobody's Perfect

The task of ergonomics is to find out and solve the physical problems arising between us and our workplaces. We deal with physical problems at work all the time, even if mostly on an unconscious level—except when we bang our heads, elbows, knees, get a headache, neckache, or other symptoms repeatedly while trying to do our jobs. Sometimes, it's because we're too large or small for the equipment we use.

We all know that people come in different sizes and proportions. A study by NASA in 1978 indicates relatively small physical differences between U.S. white and black ethnic groups in physical dimensions, and between right- and left-handed people. But it found significant differences in the various physical dimensions between whites and blacks and people of Asian origin; and large differences between the dimensions of males and females.*

The problem is, there is no such thing as an "average person"; a person average in one dimension is not likely to be average in other dimensions. Furthermore, we have differences in eyesight, hearing, sensitivity to light, and foot and hand sizes. You probably aren't actively concerned with these things at work, but the designers of your chair, desk, computer station, foot rest, communications equipment, lamps, and other elements had to decide which population groups to fit. When we consider how quickly the working population is changing with the increase in older workers, females, and diverse ethnic groups, it is no wonder many of these people feel they don't fit into the workplace.

Furthermore, because the workplace has had to be redesigned to accommodate the physically challenged, especially as a result of the Americans with Disabilities Act of 1990 (ADA), such handicaps as hearing, sight, learning, and mobility impairments all have to be addressed more frequently and by using new technologies.

Approximately 80 percent of disabled people are able to work if given the properly designed facilities and equipment. So power switches are being made more accessible, disk drives more convenient, systems adapted to the blind, screen displays modified to replace the audio signals that are inappropriate for deaf users, and programs designed that would

*The study found that the average stature of Japanese males is about 12 centimeters (approximately 4.7 inches) less than white males; average sitting height is 4 cm (1.6 inches) less than white males; average knee height is 6 cm less than white males and 8 cm less than black males; and average elbow rest height is 4 cm less than white males and 6 cm less than black males.

NASA also found that the average stature of males is 15 cm (6 inches) more than females; average crotch height is 11 cm more than females; average elbow rest height 3 cm over females; and average sleeve length 13 cm more than females. White males were also found to be significantly larger than people of oriental origin and females in chest, waist, buttocks and biceps circumferences.

let the deaf communicate with people of normal hearing over the phone.*

Biomechanics: Man as Machine

To understand how you operate at work, let's look at a still-newer discipline: biomechanics. This is the study of the human body, in mechanical terms of forces and their effects on masses, and is based on the fields of anatomy, orthopedics, and physiology. Like it or not, we're basically all machines when it comes to figuring out questions like what movements we can execute, how fast, how long, how many in repetition, and with what degree of effort or strain in order to perform a job efficiently.

Ergonomics and biomechanics help answer many basic problems in the design of your workplace. Is the work surface too high or low for you? Must you frequently sit on the front edge of your chair? Must you assume an unnatural or stretched position in order to see or reach controls, write or type, or get access to materials or parts of your work unit? Are dials and equipment controls difficult to operate or poorly labeled? Are you required to hold up your arms or hands without the help of armrests? Do you bend over to write for

*JAN, the Job Accommodation Network, is an international information network and consulting resource for accommodating persons with disabilities in the workplace. It is a service of the President's Committee on Employment of People with Disabilities, and the toll-free number in the United States is 1-800-JAN-7234. In Canada, it's 1-800-JAN-CANA.

More than eleven thousand skilled people with disabilities have been placed in competitive employment through an industry-managed program called Project With Industry (PWI). The program was initiated in 1977 by the Electronic Industries Foundation. Funded through the U.S. Department of Education, a network of sixteen affiliate offices nationwide offers free placement services to employers. Sites include Florida, Illinois, Pennsylvania, New Jersey, Delaware, California, Minnesota, Massachusetts, Nebraska, Tennessee, and the Washington metropolitan area (Maryland, District of Columbia, northern Virginia). For the number of the office nearest your location, call Carol Dunlap at 202-955-5816.

long periods at a time? Are you seated so close to the person next to you that your leg and arm room is limited? Do you stare into computer screens for long stretches of time without a break? Was any or all of your workspace designed with *you* in mind?

To explore the obvious kind of problem that occurs when job conditions and people are not properly matched, it is sometimes easier to look at an industrial setting.

> A group of laundry employees began experiencing a sudden increase in skin rashes on their forearms. There were no obvious solutions, since the work they did had been done for years with no symptoms of this type. The workers sorted laundry from tubs by lifting handfuls of it, briefly holding items up to drain, and then sorting them in bins for drying and, subsequently, for folding. Since the cleaning materials were known to be slightly allergenic, gloves were required to protect the hands of those doing the job.
>
> Armed with these facts, the ergonomist had little reason to suspect any particular cause for the increasing number of skin rashes. In the observation of the job, however, one problem was noticed that seemed to contribute to the symptoms. When working with larger pieces of laundry, the people had to hold their arms higher while draining the laundry. This allowed the water to run down their gloves and drip on their forearms. The frequency of this contact was responsible for the increasing number of rashes that were occurring.
>
> When looking for an explanation as to why this had not shown up earlier as a serious problem, the ergonomist found that the work population had changed recently and now contained a much larger number of immigrants from southeast Asia. These people, who generally were of shorter stature than the traditional employees, had to reach proportionately higher to allow the laundry to drain. This higher reach then allowed the laundry water to drip down the gloves and onto the

arms rather than just dripping from the gloves back into the sinks.

The problem, once confirmed, was relatively easy to solve. The wooden platforms that stood by the sinks were raised slightly so that the dripping was once again contained by the gloves. This example, of course, would have been very difficult to solve without the direct observation provided by an on-site study. On-site studies can also identify additional problems not noticed before.

But in an office environment, the problems are even more subtle. For example, computers are programmed to hear us in a number of ways. In many voice mail systems, the "pound" button (the # key on a touch-tone phone) is used to transfer calls or erase messages. If your voice sounds like the # tone, and most women's voices are of that higher pitch, you can expect that your message may be lost forever in computer oblivion. Some phone mail users have to ask their women callers to lower their voices, so they're not automatically hung up on by the computer. (The problem also affects male callers and callers with regional accents when any of these people speak with or use a higher pitch in their conversation.) So, even our voices must be a factor if we're to design equipment for the diverse workforce of the nineties.

Ergonomics attempts to answer questions like: Do you have to lift and carry too much weight? Must you do so while your body is bent, twisted, or stretched out? Is the work pace rapid and not under your control? Do you complain about fatigue? Does your job require one motion to be repetitively performed at high frequency? Are you required to maintain the same posture, either sitting or standing, all the time?

M. is a project manager at a major brokerage house. Part of her job is to manage the conference calls, as well as all of the special events and client meetings. She spends a good deal of her day talking on the phone while she's writing. To do this, she does what many of us tend to do.

> She cradles the phone between her ear and her shoulder by bending her head to the left and lifting her left shoulder to hold the phone in place. "The pain I've developed in my neck has become sheer agony," she tells us. "It got so bad, I tried to correct the problem by stretching my neck and turning my head in the other direction. That only made it worse. My chiropractor treated me six times and then insisted I get a headphone."
>
> Royal Wilds, M.'s chiropractor, practices in New York City and has seen dozens of cases just like this, affecting both men and women. "Cervical-dorsal problems like this are quite common," says Wilds. "When it's more protracted, that is to say in time, holding unnatural positions like this could lead to neurological and circulatory problems affecting the neck, shoulders, arms and hands. It could also lead to myofibrositis, which is the shortening and inflammation of the muscles and the tough fibrous coverings over them. Beyond that, it could also lead to a permanent problem with the alignment of the vertebral segments of the spine, and the ligaments that link them. These segments allow us to move as freely as we do in all directions, and that movement may become seriously impaired."

Dr. Wilds believes it is important for office workers to shift their positions frequently, and concentrate on good posture. Slouching tends to stretch ligaments and joints, leaving them permanently out of place.

Mainly the Strong Survive

Ergonomics helps provide practical solutions to frequent problems at work like neck and back strain. It also helps measure how much muscular strength is required to do a job without injuries; and how much cardiovascular endurance is required for heavy jobs without risks of heart attacks, strokes, high blood pressure, and elevated heart rates. These problems are often complex, since muscle strength, for example, is a complex function of inherent muscular capabilities, mechanical advantages, skill, and motivation.

Is It Really Your *Work Environment?*

Was your work environment designed for *you?* Probably not. Yet, the changing nature of the office and work requires more than ever that one's workspace be designed ergonomically for effective performance. Traditionally, an employee worked at a desk with a lot of room for various postures, movements, and activities. If it was too low, too high, or otherwise unsuitable, it was unlikely to cause great discomfort.

However, according to one survey, 85 percent of U.S. office workers say they use a personal computer or a terminal at work, up from 78 percent in 1989, and 66 percent in 1986. As a result, more of us are working in a restricted manner—often in just one posture or position—for several hours or a whole day. So, we're more susceptible to physical and mental strains from poorly designed furniture, inadequate lighting, and other conditions.

The organization with the longest-standing reputation for understanding ergonomic issues is Steelcase, Inc., a worldwide network of independently owned companies that provides office furniture and fixtures. In 1991, Steelcase did an international study of office workers, using Louis Harris Associates, and found large discrepancies in what workers say is important to them and what is true of their own office.* The largest discrepancies were:

*For a copy of the *Steelcase Worldwide Office Environment Index 1991*, contact Steelcase, Inc. in Grand Rapids, Michigan. And while you're at it, you should try to get a copy of *Office Access*, a wonderfully graphic explanation of ergonomics sponsored by Steelcase and published by the Understanding Business, a San Francisco–based division of Harper Collins Publishing.

	U.S.	*Canada*	*Japan*	*Europe*
Proper lighting	[92%] (64%)	[90%] (62%)	[87%] (55%)	[81%] (50%)
Comfortable heating and air-conditioning	[82%] (44%)	[84%] (37%)	[85%] (45%)	[72%] (37%)
Enough workspace surface	[75%] (46%)	[77%] (48%)	[75%] (30%)	[63%] (40%)
Enough privacy and quiet	[60%] (30%)	[66%] (31%)	Privacy is scarce	[56%] (25%)

[ ] = Very important
( ) = True for their office

Repetitive Strain

Repetitive strain or motion injuries are now called "cumulative trauma disorders." CTDs, such as back injuries, carpal tunnel syndrome,* tendonitis, and finger pain are causing more and more employees to see doctors and even surgeons. Worker's compensation costs are skyrocketing.**

So while injuries used to be more common in industrial settings, the office is quickly catching up. Consider the case of Claudia Deutsch, a reporter for *The New York Times.*

> In mid-January of 1992, Claudia started noticing wrist pains, but ignored her symptoms. She'd been using a keyboard for at least five years, and had no problems before. But she then changed jobs, and though she took her keyboard with her, she changed workstations. Her beat had changed, so she began doing a lot of reporting and taking notes on the keyboard as part of getting im-

*The problem that occurs when your wrist nerves get irritated, and the tendon sheaths swell, from constantly bending your hands back toward your body, something you do when typing for long stretches of time.

**Patricia Fernberg, associate editor of *Modern Office Technologies,* estimates that carpal tunnel syndrome alone costs U.S. employers over $1 billion a year.

mersed in her new beat. She found herself using the computer for more hours at a stretch than ever before.

The day she realized how bad her pains had become was when she drove her car ten blocks and had to stop. She rubbed ice up and down her arms for forty-five minutes. Says Claudia, "I was not a happy camper." She remembers that day, Saturday, January 25, very well.

She has had to stop her typing. She has to go for physical therapy several times a week, and has a whole regime of exercises she must do. She even needs to use a knob on the steering wheel of her car in order to make a turn.

Claudia's still not sure of what all the factors were that led to her condition. Was it the increased typing alone? Or the combined stress of changing jobs plus stress at home at the same time her keyboard usage went up?

Some doctors believe that the focus on ergonomics may be misguided. Dr. Peter Nathan of the Portland Hand Surgery and Rehabilitation Center in Oregon says he has conducted his own study and seen little connection between keyboard use and carpal tunnel syndrome. And Chansoo Kim, a physician who specializes in physical medicine and rehabilitation in Illinois, points out that avocational or recreational activities outside of your job could be just as responsible for repetitive strain injuries—activities like bicycling, gardening, needlework, driving, baking, tennis, writing, etc. His examination of the data also points to the seemingly higher risk factor for women who use oral contraceptives. In addition, his analyses of repetitive stress show that heredity could play an important role.

So, while some researchers concentrate on the individual differences in who gets repetitive stress injuries and who doesn't, and how much the office is to blame, we are left with several choices. We could concentrate only on those activities outside the office that contribute to our physical ailments. And we can go on and design offices, buy equipment, and perform our work as if ergonomics didn't matter. Or we could acknowledge that perhaps we ought to alter our tasks and our surroundings to our bodies wherever they may be to prevent the stress and pain that could come from the way we lean, move, see, work, and sit.

Since the keyboard seems to be a major culprit, here are some tips for those who must use a keyboard a good part of the day, gathered by Jane Brody, reporter for *The New York Times*, and Pam Black, writing for *Business Week:*

- While you are at your keyboard, keep your back angled a few degrees back, to widen the angle between your torso and thighs. This will increase blood flow and reduce the compression of your spine. Make sure your arms are relaxed and loose at your sides, all the while keeping your forearms and hands parallel to the floor. Your thighs should be at a right angle to your torso, and your knees at a right angle to your thighs.
- Your chair should have a back rest that fits the curve of your lower back. If it doesn't, slip a cushion or foam padding behind the curve in your lower back. Your seat should slope slightly forward, in order to move some of the pressure off your spine by pushing it down into your thighs and feet. And use a seat cushion that curves downward in front to ease the pressure on your thighs.
- Also, do periodic exercises with your hands. For example, massage them inside and out with your thumbs and fingers. Gently, bend your wrists back for several seconds at a time. Repeatedly, clench and release your fist, fanning out your fingers when you open your hand.
- Always work the keyboard with your elbow at a 90 degree (right) angle, and keep your wrists straight. Do not arch them up or bend them down. Straight wrists are a critical factor in preventing injuries. Stowable racks or books can lift your keyboard if it is too low now. And

using a wrist rest to support your wrists is especially helpful during typing breaks. More keyboards are now being sold with wrist rests, but you can also buy a rest separately and attach it to your existing keyboard.

• If you speak on the telephone often while writing, working, or using your keyboard, buy a headset for your phone. This will significantly ease the stress on your neck and upper body. Headsets are sold at office supply and electronics stores.

• And, remember, always take frequent breaks. And fidget. Yes, fidget. Despite what teachers and parents told you, sitting still for a long period of time is *not* good for you. Occasionally shift your weight and stretch your arms and legs. But good posture *is* important.

An ergonomically designed workplace should reduce your muscle fatigue and chances of developing musculoskeletal diseases like tendon and ligament inflammation and pain. For example, it has been shown that workstations should not require your hands to reach frequently or be held for lengthy periods above your shoulder height, or you will risk the development of degenerative tendonitis.

Some entrepreneurial firms are exploring ways to develop new products and services to make the office more hospitable. Keyboards and chairs are the two most active areas of innovation. Industrial Innovations, Inc., of Scottsdale, Arizona, has developed the Datahand keyboard. No longer requiring you to take on the demeaning position of a begging dog while at your keyboard, Datahand is designed for the natural position of the hands (keys line up with the fingers), and your hands needn't move. Dozens of other keyboard designs are in the works and slated to roll off assembly lines in the next few years.

Sausalito, California, engineer Hector Serber invented Ergomax. It is a chair with seven independently adjustable parts.

The back rest can swing forward to be used as a chest support. The foot rest adjusts upward to be used as a kneeling support. And the cushions are made to insure proper skin surface circulation and comfort. It sells for $1,000. All the top office furniture makers and dozens of smaller entrepreneurs are researching and designing a whole new generation of office chairs. Some experts believe that office chairs should no longer be viewed as furniture, but as a tool and a factor in orthopedic medicine.

A firm called Great Performance in Beaverton, Oregon, publishes an *Occupational Health and Safety* mail order catalog. In it they offer Lifeguard, a computer software program that prevents computer-related injuries by keeping track of your keyboard keystroke activity. It will warn you with friendly on-screen reminders when you are doing something in a way or at a pace that will cause damage to yourself. Lifeguard comes complete with illustrated exercises and diagrams on how to adjust office furniture for good posture.

And furniture makers are certainly beginning to recognize all aspects of the ergonomic marketplace. One desk maker (Haworth, in Holland, Michigan) is producing a desk that beeps periodically to remind its user to push a button that alters the desk height to another of its sixteen positions, thus getting you to shift your own position every forty-five minutes or so. This little bit of physical therapy would run you $3,500 to $4,200.

We have one client who had a desk designed especially for him—he occasionally likes to *stand* and work, so he has one desk for sitting at and one desk for standing at. Not a bad idea!

8 *Say A-h-h-h: Diseases in the Office*

When most people worked in factories and on farms, the primary hazards were from machinery. Malfunctioning machines, and those with poor safeguards for workers, along with worker carelessness, were the main known threats to physical well-being. Over time, we also came to recognize other dangers: chemicals, for example.

Now that much of our work has shifted away from field and factory, we tend to think that workplace hazards are a thing of the past. After all, with no big, dangerous machines and with no apparent abundance of chemical substances, the office seems to be about as risk-free as any workplace could be. Oh, yeah.

Sick Building Syndrome

In her book, *The Nontoxic Home and Office*, author Debra Lynn Dadd lists some of the health hazards to be found in the typical office:

- Adhesives
- Artificial lighting

- Asbestos
- Cleaning products
- Coffee
- Computers
- Copying machines
- Electromagnetic fields
- Magic markers
- Paper and paper goods
- Pesticides
- Polystyrene cups
- Radon
- Synthetic fibers
- Tobacco smoke
- Typewriter correction fluids
- Urea-formaldehyde foam insulation
- Ventilation systems

Is it any wonder that we have that phenomenon known as "sick building syndrome"? The latest estimate is that between 800,000 and 1,200,000 commercial buildings in the United States can be sick buildings. Workers in these buildings complain about headaches, respiratory infections, rashes, nausea, and other ailments in numbers sufficient to indicate that there are environmental problems.

The New York Times recently reported on one building in Albany, New York, that could serve as a case study for sick building syndrome. The building, which houses the state's Department of Taxation and Finance, has been the scene of a series of health problems that could rival the Biblical seven plagues.

Over two years ago, there was an outbreak of skin rashes among people working in the building. In the most recent event, dozens of people suddenly became dizzy, began vomiting, and even fainted. During the two-year period, there were many instances of employees getting headaches, feeling disoriented, and exhibiting other symptoms of physical discomfort.

Some observers, inclined to question how real these problems are, speak of causes such as mass hysteria and malingering. Others may believe that people who work for the tax collector are only getting what they deserve.

The truth is, however, that the building is an unhealthy place. During an inspection, it was discovered that the ventilation system was so clogged with dirt and debris that it wasn't ventilating some parts of the building at all. It was also discovered that a pesticide used in the building was making some people sick.

At the first conference of the National Coalition on Indoor Air Quality in May 1992, the case was raised of an insurance company that moved some of its workers into a building on Myrtle Street in Hartford, Connecticut. Soon the workers complained of sinus infections, sore throats, runny noses, skin irritations, and headaches. Over the seven-year period that followed, some claimed they developed asthmatic bronchitis and other chronic diseases. Even though the company finally moved, 140 employees filed suit against the building's owner, making it Connecticut's first class-action suit of its kind. There are (as of mid-1992) a number of states that have proposed or passed legislation to deal with this issue, including California, Michigan, Missouri, New Jersey, New York, Oklahoma, and Oregon. Congress is also working on efforts to research and curb the problem.

Two scientists at Georgia Tech conducted an experiment in which they pumped more air into a problem building. In this building, there were high levels of carbon dioxide, which makes people lethargic, and formaldehyde and volatile organic compounds (VOCs), both of which can cause allergic reactions. The increased volume of fresh air reduced the pollutants and the health complaints.

A study published a few years ago in the *Journal of the*

American Medical Association found that people who work in modern buildings come down with respiratory ailments 45 percent more often than people in older buildings. Another study done for the National Institute for Occupational Safety and Health (NIOSH) said the single most important factor in sick building syndrome was poor ventilation—a characteristic of modern buildings. Doctors are reporting increasingly severe cases of immune-related diseases caused by indoor environmental pollutants. Asthma, for example, has risen in incidence 58 percent since 1970.

Aside from physical illness, psychological problems are also being reported. Psychologist Russell Bauer of the University of Florida in Gainesville directed a study in which he and his associates did health and psychological surveys of ill people. Twenty-seven worked in a building where they complained of sick building syndrome symptoms for as long as two years. Fifty-eight worked in the same building but did not suffer from SBS. And twenty-six, who served as the control group, worked for the same company but in a different building, where no SBS problems were reported.

The two groups that worked in the same building, including those that had no physical complaints, scored much higher than the control group in the other building on many areas of psychological stress. These included depression, paranoia, defensiveness, anxiety, confusion and resentment, and distrust of authority.

(Employees that complained of SBS tended to smoke cigarettes much more than those who didn't, so there is a possibility that smoking may make you more sensitive to airborne contaminants in poorly ventilated buildings. Or smoking might impair your immune system enough to make you more vulnerable to SBS.)

Another contributor to sick building syndrome is radon, the gas that, according to current scientific opinion, may well be more hazardous than asbestos or pesticides. And recent research suggests that copiers and laser printers, ever more present in modern offices, emit ozone in amounts sufficient to cause respiratory ailments, nausea, headaches, and premature aging of skin. High-intensity quartz-halogen bulbs, increasingly used in offices, may also be hazardous. There are some preliminary indications that the light from unfiltered bulbs could cause skin cancer.

Interestingly, a study done in England seems to show that air-conditioning, no matter how good a job it does of ventilation, is inferior to natural ventilation. Even naturally ventilated buildings with high concentrations of airborne fungi, bacteria, and other contaminants had lower frequencies of sick building syndrome symptoms than good-circulation air-conditioned buildings. Maybe this has something to do with the lighting conditions (see chapter 1). In any event, windows (especially ones that open) may be your best solution to depression, allergies, and fatigue.

Electromagnetic Fields

The growing presence of electronic equipment in offices is also resulting in a significant growth in complaints about health. This equipment generates extremely low-frequency electromagnetic fields (ELF), usually at 60 cycles per second or 60 hertz. Evidence does not now absolutely link ELF to illnesses or deterioration of health. But there is a developing body of scientific research that raises questions about the relationship of ELF—particularly constant exposure to it—to human health.

Some of this research is contradictory. There are studies, for example, that show a relationship between exposure to power lines and childhood cancer. Other studies do not show such a link.

A study issued in 1989 by the Office of Technology Assessment, the U.S. Congress's research arm, said that "emerging evidence no longer allows one to categorically assert that there are no risks, but it does not provide a basis for asserting that there is a significant risk." What? This is obviously Washington's way of coming down as firmly as possible on both sides.

As more people become exposed to ELF in the workplace, more research is being done. Perhaps we will know fairly soon whether there really is a problem, and how much of a problem it may be. Meanwhile, concern mounts.

A couple of recent studies conclude that electromagnetic fields might affect the relationship between metal ions (such as calcium) and some proteins in the body. Changed calcium levels in the body could have either beneficial or harmful effects. For example, calcium deficits in aging women could lead to osteoporosis. So this is one area that is certainly deserving of more research.

A Department of Energy study found that men had slower motor responses and heart rates, and altered brain wave patterns, after being exposed to electromagnetic fields of the sort that surround power lines.

Another study by Johns Hopkins University of over fifty thousand males showed a high risk of leukemia, especially for younger men who work extensively around ELF. And still other studies have linked brain tumors and ELF exposure. Some researchers have found that low-energy electromagnetic fields bring about changes in an enzyme, ornithine decarbonoxylase, that plays a role in the growth and reproduction of human cells, and that also appears in cancer cells.

In 1987, Dr. Robert O. Becker told the House Subcommittee on Water and Power Resources that specific organs in every living creature serve to sense changes in the earth's magnetic field and alter the creature's behavior accordingly. Increase in electromagnetic energy in an organism's environment can affect the organs' response, and possibly result in diseases and abnormalities.

(Dr. Joseph Kirschvink, a geobiologist at Cal Tech, claims that human brain cells contain crystals of magnetite, a highly magnetic mineral. Magnetite crystals, according to Kirschvink, enable organisms from bees to whales to navigate by using the earth's magnetic field. If he is correct, and very few scientists so far agree with him, these crystals could account for at least some of the influence of electromagnetic fields on human health.)

Almost everybody these days works with a computer—a word processor, a PC, a workstation, whatever. All of these give off nonionizing radiation (NIR). The Environmental Protection Agency discontinued some years ago its research and standards program for NIR. Meanwhile, there is growing concern that all these forms of electromagnetic radiation in the workplace can seriously affect human reproduction.

A study done several years ago by Kaiser-Permanente showed that women who used video display terminals (VDTs) for more than twenty hours per week had almost twice as many miscarriages as women not using the terminals.

IBM is sufficiently concerned that since 1989 it has offered VDTs with reduced emission of electromagnetic fields. And there are protective screens now available on the market to help shield against the exposure. If you do not already have one, you would be wise to get one.

And it's not just low-frequency radiation. In 1990, Denmark's Telecommunications Authority warned users of mobile telephones, which emit high-frequency microwaves, that these could cause brain damage. Extended use, said the Authority, was equivalent to sticking one's head into a microwave oven. There's an image for you.

Chemical Soup

The modern office is also full of chemical hazards. Office furniture, for example, contains chemicals that can cause headaches, allergic reactions, and respiratory ailments—all, of course, exacerbated by the poor ventilation and circulation characteristics of modern buildings. These chemicals include vinyl chloride and polychlorinated biphenyls (PCBs), which may also be carcinogenic. These and similar chemicals are known to affect human DNA, producing hormonal alteration or aberrations in the immune system, which can lead to cancer.

PCBs, for example, may be a significant factor in the breast cancer epidemic in the United States. (We use the term "epidemic" because latest statistics indicate the incidence of breast cancer has increased significantly over the last two decades. It is now estimated that one in nine American women will get breast cancer.) Researchers have found that tissues of women with breast cancer show over 50 percent higher concentrations of PCBs than normal. Although now banned, PCBs were used extensively in the past in office furniture and equipment.

Formaldehyde, another commonly used chemical, has lately become as notorious as asbestos. It's pervasive in the workplace. It is used in insulation, carpets, paint and wallpaper, fabric dyes, floor and furniture wax, plaster, plywood—just to name a few. Formaldehyde can cause respiratory and eye problems, skin problems, and—in high quantities—cancer.

The dangers some chemicals in the workplace represent to women of child-bearing age have long been a concern to society. Government agencies such as EPA and OSHA have devoted considerable efforts to protecting such women. What has been given relatively little attention, until very recently, is the dangers in these chemicals for men.

Young men exposed to toxic substances may be just as likely as women to cause defects in their children. As mentioned in chapter 3, studies have shown that male gonads and sperm can be adversely affected by substances ranging from industrial chemicals to tobacco. In fact, the Supreme Court has recently ruled that fetal protection policies that apply only to women are unconstitutional.

Although cigarette smoking is now limited or forbidden in many offices, by law or company decision, there's still a lot of it around. Indeed, there is even a smokers' rights movement that is trying to overturn rules against smoking. It is increasingly evident, however, that passive smoking—inhaling the smoke from other people's cigarettes—is a genuine health hazard. In the first such case, an Australian court ruled in 1991 that there was conclusive evidence that passive smoking causes lung cancer, asthma, and respiratory problems. Many observers feel this case raises serious liability questions for employers who permit smoking on their premises.

The cigarette issue raises other interesting questions. Why do only some people who smoke cigarettes get lung cancer? Why not, if tobacco is as toxic as most scientists and doctors believe, all smokers?

Our Genetic Destiny

One answer, obviously, is that biology is not just statistical; it's also individual. Although we're all human, and we have a common biology, we also differ from each other. Some people have adverse reactions to everyday items—perfume, for example, or even newspapers. Just a whiff of perfume can cause headaches, nausea, shortness of breath, fatigue, and even depression in such people. These people (perhaps up to 15 percent of Americans) are hypersensitive, and they are victims of today's hot medical topic—environmental illness.

Traditionally, there have been three primary causes of illness: infections, environment, and genetic makeup. (There is a fourth—iatrogenesis, or illness caused by medical treatment—but the health care establishment doesn't like to talk about that one.) It is becoming increasingly apparent that the most important factor is genetic. So, hypersensitivity, the basis of environmental illness, is determined by the individual's genes.

Many health practitioners dismiss hypersensitivity as psychosomatic because traditional tests often fail to find any physical cause for the symptoms. One well-known immunologist, for example, said that "chemical hypersensitivity constitutes a belief and not a disease." But a growing number of people, including some in medicine and science, suspect that the causes of hypersensitivity just haven't been discovered yet.

Another genetic-related factor has to do with race. Chinese men, for example, seem to be more sensitive than Caucasian men to one of the most widely used heart drugs, propranolol. The Chinese need only half as much of it as Caucasians. Doctors are warned to take into account the relationship between ethnicity and reaction to medication. And it is widely known that, in general, Asians have a lower tolerance for alcohol than do Caucasians.

As the racial, gender, and age makeup of the workforce changes, the differences of reaction and susceptibility among the different population segments will have to be taken more seriously. The Bureau of Labor Statistics (BLS) projects that 47 percent of the labor force by the year 2000 will be female, up from 45 percent in 1990. Blacks are expected to be 12 percent of the labor force in 2000, up from 11 percent in 1990. And Hispanic workers will constitute 10 percent in 2000, compared with 7 percent in 1990. As for age, the percentage of the workforce under thirty-five is expected to decline significantly, while the over-thirty-fives will increase, particularly those forty-five and older, as the Baby Boom ages. The forty-five–to–fifty-four age group, now 16 percent of the workforce, will be 22 percent in 2000.

Yet another and often overlooked biological factor in the workplace is left-handedness. In his book, *The Left-Hander Syndrome*, Stanley Coon points out that the approximately 10 percent of people who are left-handed have a shorter life expectancy than right-handers—approximately nine years shorter, according to him. Coon says that left-handed people suffer more accidents because they live in a right-handed world. Certainly, most, if not all, workplaces are designed as if there were no left-handers working there.

All of this demonstrates the need for what Laurence Foss and Kenneth Rothenberg, in their book, *The Second Medical Revolution*, called a shift to an "infomedical" health model that focuses primarily on people as individuals rather than on their afflictions. As we move toward a more holistic and a more individual treatment of people and their ills, taking into account how each person interacts with his or her work environment will become increasingly important and useful.

In the future, managers and owners of businesses will have to give more consideration to this approach. The Americans with Disabilities Act of 1990, in the opinion of some medical and legal authorities, could be interpreted as defining a genetic flaw as a disability. Thus a genetic susceptibility to a chemical could be a disability, and poor treatment of any individual with that problem can be interpreted as discrimination. Some lawyers even argue that, under the ADA, rejecting a job applicant because of a preemployment physical can be seen as discrimination against the disabled.

We are more and more obsessed with our health. And we are all dependent on work, not only for money but also for our sense of identity and worth. As the two, health and work, become more closely intertwined, our demands for a healthy workplace will intensify.

But it's not all bad news. Some research done for NASA a few years ago showed that several houseplants filter organic pollutants from the air. There is evidence that plants such as the spider plant can significantly reduce amounts of benzene, trichloroethylene, and formaldehyde—all contributors to sick building syndrome. Three space scientists (from the United States, Russia, and Canada) who have spent years researching how to set up space colonies say a forest-like environment in the office may be a powerful weapon against SBS. If there is too much formaldehyde from cheap furniture in your office, they recommend Boston ferns and chrysanthemums to soak up the fumes. And if you're exposed to too much benzene from tobacco smoke or carpets, try surrounding your desk with gerbera daisies and peace lilies.

9 *Cybernetics: The Human/ Machine Interface*

For decades, science fiction has been depicting an evolutionary process in which humans are increasingly linked with smart machines. Robots have been envisioned as man-machines, and TV's "Six Million Dollar Man," the "bionic" wonder, was a man reconstructed with incredibly strong artificial body parts. Now along come *Terminator*, and *Terminator II*, in which Arnold Schwarzenegger plays the ultimate fusing of man and machine.

Medicine has long been exploring the role of linking humans to mechanical life processes, whether we go back to the introduction of the iron lung and kidney dialysis, and even the pacemaker, or forward to the micro-miniaturization of electronic implants that can help regenerate damaged nerves or restore vital organs.

In the modern world, we have rapidly become accustomed to the human-machine interactions that occur every day as we work with ever more "intelligent" computer systems in our daily business. Right now, these machines are separate from us bodily. But quickly approaching is virtual reality—the ability of machines to mimic human senses of sight, sound, smell, taste, and touch to the point where, with the proper attire (like electronically linked gloves and eyewear), we can actually experience being somewhere we aren't. We can even interact with a distant or imagined environment.

Through virtual reality, for example, Japanese consumers in a furniture showroom are now projecting themselves into their own kitchens and exploring how various kinds of cabinets would look and work in their homes. Biologists are already able to "step inside" molecules and take a look around. And architects can "walk into" buildings planned but not yet built, so they can test the aesthetics and workability of their blueprints. In video games of the future, the player will actually be *in* the game.

This onward push toward the fusing of humans and machines is a centuries-old process. And that leads us to look at where we have come in the past few decades, and how the way we work will continue to be affected.

The New Science of Mechanical "Feedback" Systems

Following World War II, a group of scientists and engineers, led by mathematician Norbert Wiener, began research that linked machine control and human communications. This fusing of man and machine became known as cybernetics.

Cybernetics appealed to researchers across the board: sociologists, psychologists, political scientists, biologists, and anthropologists. It was the dawning of an era that was to place the man/machine interface at the forefront of visions of the future, whether that future described medicine, war, the economy, the home, or the office.

Questions began soon after the new science was born. Would cybernetics be used to help us understand or better the human condition? Or would it lead to dehumanization, invasions of privacy, unwanted control of our activities, and the establishment of "Big Brother"? The concept of people as machines, with all the appropriate and measurable feedback systems that were coming to fuel modern technology, was a difficult one for many to accept. Prominent social scientists, such as anthropologist Margaret Mead and psychiatrist Erik Erikson, claimed that there were aspects of human brains and activities that did not fit neatly into cybernetic systems.

Steven Heims, a physicist turned science historian, raised many questions in his book, *The Cybernetics Group*. If humans are not machines, what are we? Maybe the fact that our behaviors cannot be totally programmed, as that of machines can, just means that we are machines that cannot understand ourselves. And biomedical engineering, which has provided us with artificial (mechanized and computerized) limbs and organs, would seem to confirm that much of our bodies are, in fact, machine parts.

Within the office, cybernetics has taken over. Questions such as those noted above are generally absent from discussion. It has just been taken for granted that humans and machines make a happy marriage, that from this marriage will come wonderful and productive goods and services, and we shall live happily ever after. Read on.

According to *The New York Times*, computer-related injuries now account for almost 40 percent of workers' compensation costs. Health expert Jane Brody, writing in the *Times*, says that hundreds of thousands of office workers are disabled each year because of disorders related to working with computers. The computers, in effect, are making today's office the equivalent of yesterday's factories. But this is only the surface of what is in reality a far deeper problem.

Human Factors

When the U.S. Air Force began using the F-16 in the early 1970s, it was the most advanced fighter plane in the world. It could perform without structural damage at a remarkable 12 Gs—twelve times the force of gravity. But the best pilots with the finest training and the newest G suits will black out at 9 Gs. Consequently, in the years following the introduction of the F-16, several dozen pilots died in training accidents.

That 3 G difference between the maximum performance of the plane and that of the pilot is what psychologists call "human factors"—the gap between human and machine capabilities.

The F-16 example is an extreme one. But it is by no means the only one, although perhaps more deadly than most. As technology becomes more pervasive in the workplace, the problem of human factors is one that organizations as well as people increasingly face. The great cause for concern is that few managers have even heard of the problem, let alone thought of how to deal with it.

The extraordinary—perhaps even revolutionary—development of new workplace technologies has generated reactions ranging from wild-eyed optimism to a revived Luddism. The optimistic champions of technology predict a new Golden Age, with machines liberating us from drudgery and danger.

The neo-Luddites (whose attitude can perhaps best be summed up by journalist I. F. Stone's wry remark after trying to learn how to use a word processor: "The only thing God didn't do to Job was give him a computer!") see serious economic and social problems ahead for us.

And that's because not enough attention has been given to the interaction of people and technology—and, most specifically, *how to improve it.* It must be remembered that the major, if not the only, purpose of workplace technology is to improve human productivity. The purpose of focusing on human factors is to maximize productivity gains—and to minimize, if not eliminate, the obstacles to such gains.

The year 1986 saw two striking technological calamities—Chernobyl and *Challenger.* Both came in highly advanced technologies, programs with long success records. Analysis by both American and Russian scientists resulted in startlingly similar conclusions: the cause was not poor technology but poor technology management.

The U.S. Armed Forces are bedeviled by human factors problems. It has become increasingly apparent that the war against Iraq was not the complete technological success we first thought it to be. Indeed, the more we learn about it, the more we see that our own high-tech weapons may have been a more effective enemy than the Iraqis.

High-tech weapons require skills, for both operation and maintenance, that too many young people—even high school graduates—lack. The attempt to develop the automated, robotized "factory of the future" has run into a serious people problem: most workers lack the skills to run and maintain the machines. At General Motors, it has been reported, 15 percent of hourly workers are illiterate. Often, because of union rules, workers retrained for the new skills are those with seniority, many of whom are close to retirement.

There are many analysts who now believe that, contrary to earlier expectations, the increasing use of high tech in manufacturing will require more, not less, worker involvement—and higher levels of skills.

The service sector, which now employs more than 70 percent of workers, has been far from successful in benefiting from workplace technology. It has been estimated that between $600 billion and $800 billion was spent on computers for the service sector in the eighties. Yet service sector productivity grew by less than 1 percent per year in that decade. High tech, according to some economists, now accounts for more than one-third of all capital equipment outlays, and the ratio of high-tech capital spending per worker has more than doubled in the last ten years. The presumption is that such high-tech investment guarantees a productivity payback, although it has yet to be proved.

For business as a whole, according to one recent study, data processing budgets increased an average 12 percent a year over the last ten years while productivity increases averaged less than 2 percent. Maybe this was because managers believed the human factors gap was bridgeable through ordinary efforts in education and training. We now know this is a mistaken belief.

A sidebar to this, which would be humorous if it weren't so costly, is what SBT, a Sausalito software company, calls "futzing." In *Newsweek*, SBT reported that as much as 2 percent of America's gross domestic product is "futzed away" by those who spend too much time "font futzing" (prettying up copy with fancy typefaces), "office blocking" (writing the same thing over and over because tinkering on the word processor is so easy), "spread-webbing" (getting caught up in overly elaborate spreadsheets), endlessly polishing up charts, graphs, and other displays, and so on.

Getting Reoriented

Perhaps above all, the new technologies are disorienting. They have a profound effect on our biologically and culturally ingrained perceptions of time and space. People who use computers regularly find themselves adapting their own biological rhythms to those of the computer. Everything gets speeded up; adrenaline levels stay high.

Advances in communications technology now mean that everything happens instantaneously everywhere. Distance has vanished. Traditional rhythms of day and night, work and leisure, become increasingly irrelevant as the new technologies make possible—indeed, require—the twenty-four-hour day and the seven-day week. Beepers, cellular phones, PCs—these and other developments mean that we will always be linked closely to the job, unobstructed by time or space.

The effects of the disorientation these changes cause are seen in many ways. Most striking is the growing perception by more and more of us that we have less time than before. In fact, some studies show that most people actually have *more* time. The apparent contradictions show the befuddlement created largely by the impacts of new technologies on our sense of time and place.

Also changed are our perceptions of other people. Just as machines are accessible twenty-four hours a day, seven days a week, we seem to expect people to be always and evenly accessible. And people who are slow to respond to electronic communications—or faxes, for that matter—seem more and more to us to be painfully slow or even sluggish.

The noted scientist Robert W. Lucky of Bell Labs writes of the changes technology has brought about in the relationship between people and information. Technology has created a sort of supply side ergonomics, in which the availability of information drives the relationship now. Lucky notes how he and his colleagues are beginning to dread vacations—being out of touch, perhaps missing something in the massive and continous flow of information, causing them to be at a disadvantage. This seems to be accompanied by a fear that getting back into the information flow is like getting onto a California freeway during rush hour.

Harvard Business School professor Shoshana Zuboff has studied the disorienting effects of the new technologies in the workplace. In her masterful book, *In the Age of the Smart Machine*, she says that computerization has resulted in "an essential change" in how workers see their world and their place in it. For one thing, the new machines become intermediaries between them and their work, so they become more remote from that work, which in turn becomes more abstract and less real to them. Instinctive work processes have to be forgotten and new, alien ones learned.

What results is a discomfort and a sense of unreality that is not unlike one's first exposure to a form of sensory deprivation.

This psychological disorientation tends to have physical side effects. Workers (at all levels) feel cast adrift and often get physically ill or develop disabling symptoms. Eye strain and backache are two such symptoms, and they often seem to have no direct or specific cause.

One of the more novel side effects was reported on recently by Stewart Brand in his magazine *Whole Earth Review.* Military pilots who spend a lot of time training on advanced airplane simulators are developing a condition known as "simulator sickness"—visual distortions and physical disorientation apparently resulting from what one psychologist calls "cue conflict." That is, the simulation is too close to reality, and the mind apparently needs a clear distinction between the two.

If you are a parent whose children are devotees of video games, you may want to watch out for simulator sickness to occur as the games become less cartoon-like and more realistic.

What gives urgency to the human factors issues is the continuing, almost relentless advance of technology. As wide as the human factors gap is now, it will get wider. To put it plainly, technology is improving faster than we are.

Info "Overload"

The overwhelming human factors problem is that we have technology that enables us to gather far more information than we as humans are capable of processing. This is generally called "information overload." But that term may be misleading, because it implies that what we need is less information. What we really need, however, is *better* information—that is, information that is more useful, more usable, and more sensitive to how we process information.

We had a food company as a client. The marketing executives used to get their market reports monthly. In the month-long interval between reports, these marketing people could analyze the data and develop the most appropriate tactics to improve or maintain deliveries and market share. Then these managers started getting their reports weekly, and there wasn't enough time to process the information properly. Before long, the reports will be available daily, an eventuality that fills them with dread, especially as the company downsizes and the people left are responsible for picking up the work load of those who were let go.

The key point here is not that there was an increase in information, but that the time factor changed. As a consequence, the information is becoming less and less manageable.

Market researchers tell us that as the number of advertisements you are exposed to goes up, your ability to recall ads goes down. According to the research firm Sami/Burke, the recall rate declined from 24 percent in 1979 to 21 percent in 1988.

One of the most advanced commercial airliners is the Airbus A320. The French pilots' union says the plane's technology is too much for the people who fly it. According to one pilot, the plane "gives out far too much information."

Author Stan Lee coined the phrase "negative information," which he defined as information that causes us to know *less* than we did before. "It subtracts from one's store of knowledge and wisdom," he wrote. "The more negative information one is exposed to, the less one knows."

In the case of the food company, what had been valuable marketing information is in the process of becoming negative information because new systems were installed without taking into account human factors.

Datamation magazine reports that the amount of information electronic engineers have to learn is already too much and is increasing at an accelerating pace. If even the most conservative estimates of information increases by the year

2000 are correct, it will be humanly impossible to learn it all.

It was recently reported that of the vast amount of data gathered by NASA in space exploration in the last two decades, only 10 percent has been studied by scientists. And only 1 percent has been analyzed closely. The problem is expected to worsen: the proposed space station will send down 10 trillion bits every day, the equivalent of five thousand laser disks per day. (A laser disk can contain the equivalent of a complete set of encyclopedias.) And the proposed earth science satellites NASA expects to have in place by 2005 will send back 2 trillion bits a day—more earth science information in one day than was gathered in all previous human history.

There are now more than forty thousand scientific journals in the world, and more are being started every day. Some experts claim that the amount of scientific literature doubles every ten to fifteen years.

Over the last decade, we were told over and over again that we were in the midst of an information revolution. There are reasons to doubt the accuracy of that statement. A true revolution requires more than what we now have: the technology. It requires a corresponding improvement in the ability of people to match the capabilities of the technology, and that has not come about.

There have been some accommodations by people to the increases in the flow and magnitude of information. One such accommodation is what has been called "nuggetizing"—we boil down the vast amount of information to concepts or statements that can be easily grasped and absorbed. Thus, information is packaged, and the more easily accessible the package the better. What results is a kind of brittle world in which all room for doubt, debate, speculation and analysis is displaced by simple statements. Witness the political campaigns of recent years.

Decision Making

All of this has powerful negative impacts on decision making. Managers fearful of being held accountable will practice defensive management, just as doctors afraid of malpractice suits will practice defensive medicine. Growing concern about data security contributes to the institutional sclerosis by encouraging a desire to hoard rather than distribute information.

People also tend to surrender to the flood of information by allowing machines to replace them as decision makers. In 1988, when the U.S. cruiser *Vincennes* shot down an Iranian civilian airliner on the mistaken assumption that it was a military plane, nobody questioned the decisions made by the Aegis system. A seaman with a pair of binoculars could have identified the plane and prevented the tragic mistake. But the great volume of information intimidated the humans, and they let the machines decide.

As we become inundated with data, we dawdle or delay, we abdicate, or we make decisions that we ourselves are increasingly dubious about. Morale plummets, and what may be an instinctive fear of technology develops into the newest occupational illness, "techno-stress."

Many people now speak openly and frequently of their disappointment with automation and computers. The gist of their complaints is that the technology has not lived up to its potential. But that is not the problem. The technology has lived up to its potential; indeed, it has exceeded expectations. The real problem is that we did not anticipate, and still do not fully appreciate, human factors.

It is first necessary to admit that there is not yet a satisfactory substitute for human intelligence. It is then necessary to examine what technology can do to provide support for our judgment—or how it might interfere. Only then can we have a proper context for the purchase and use of technology intended for the management of resources, people, and organizations.

The Blending of Human and Machine

Many people are optimistic that the current situation will change for the better. Sociobiologist Edward O. Wilson says that the revolution in information technology will ultimately enable us to transcend our human biological constraints on information storage and transmission. The implication is that we will somehow become more like computers, finding ways to expand our memories, thinking faster and on a larger scale, perhaps even sending and receiving information at incredible speeds.

There are others who believe that the machines will become more like us. Science fiction movies show us machines such as Hal and R2D2 that have human flaws and foibles as well as intelligence.

Denis Gorges, a psychiatrist, patented a machine some years ago that utilizes sound, light, and electromagnetic impulses to relax people. Imagine, if you will, a computer that does the same thing. Going into a computer store could be like going into a pet store to buy a puppy. Each computer would be trying to appeal to you, trying to make you like it enough to take it home. Doesn't it make you wonder about the eventual possibility of seduction by a computer?

Computers even now can make people whole. Electronic implants can enable physically and mentally handicapped people to move and communicate. Versions of these could eventually enable musically untalented people to bypass their own inner impediments to musical expression and sing or play an instrument with great skill.

The longer we co-exist with the computer, the more it seems to define and shape the world for us. Some scientists now see the universe as a vast software program. Everything in such a universe is seen in the context of information processing. These scientists, for example, say that the difference between death and life is the ability to process information.

All this may appear to be an effort to come to grips with the inevitable. On another level, however, it demonstrates that humans and machines are moving toward a mutual accommodation that will ultimately require a change in human nature.

10 | *The Stress Mess*

If you think the tension you experience at the office is unique to you, think again.

The National Council on Compensation Insurance says there has been a large and steady rise in workers' compensation claims that are based not on physical injury but on psychological damage. Stress now accounts for about 14 percent of occupational disease claims. It's been suggested that one in four of us suffers from some kind of mental health problem caused by stress on the job, at an annual cost of maybe $150 billion.*

How much of this is "real," in the sense of mind-body damage, and how much of this is "fabricated," in the sense of ordinary mental challenge?

*In 1990 alone, stress claims in California led to $380 million paid to employed and unemployed workers claiming sleeping disorders, headaches, backaches, and other stress-related problems. In their eagerness, some doctors (and lawyers) have even advertised to solicit those who believe they have suffered illness as a result of stress. California's governor Pete Wilson has stated that Workers' Compensation has become such a major cost of doing business in California that it discourages companies from expanding in or moving to California and creating jobs.

To disprove a stress claim, employers generally have to show that the worker's illness was caused by factors outside the job. This often means the expense of collecting a body of information on the worker's personal life, such as smoking, drinking, and eating habits, family health, and psychological profile.

William Thiele, president of the Continental Corp. in New York, suggests there is a much greater need for measurable medical standards for psychological injury. As you might expect, insurance companies believe that many stress claims have no general medical basis. Still, everyone admits that emotional injury is truly tough to prove, because the link between the mental and the physical is sometimes so subtle.

When a person is under stress, the sympathetic nervous system produces adrenaline-like chemicals that affect various sites in the body—sites that affect blood flow and, therefore, heart rate, muscles, blood vessels, and lungs. This leads to a response in the system called "fight or flight," a primitive way of coping with the stress. But, being civilized and under the control of a variety of forces, we are not always capable of punching someone or running away. The effects of not doing either include, but are not limited to, rapid heartbeat, muscle tremors, dry mouth, sweating, nausea, digestive problems, and body aches.

Some experts believe that the reason for the rise in stress-related injuries is the growth of the service sector. The thinking is that when the work is more mental than manual, the injuries will be, too.

Researchers at the cardiovascular center at New York Hospital–Cornell Medical Center, under the direction of cardiologist Thomas Pickering, have pioneered in trying to measure the physical damage done by stress in the office. They have attached blood pressure and pulse-measuring devices to a wide range of workers, from stockbrokers to secretaries, in order to record mind/body relationships at times of mental stress. No surprises here. Bad professional news, rapid-fire demands, interruptions, unexpected arrivals, internal back-biting and finger pointing, tight schedules, added work-loads—all sent blood pressure and pulse up the chart. Yet, while studies have proved that stress has long-term effects on the arteries of primates, this has not been as easy to show in humans.

The Cornell studies showed something else—that *a woman's blood pressure peaks when she's anxious; a man's pressure peaks when he's angry.*

A 1989 Gallup Poll, commissioned by the New York Business Group on Health, estimated that each employee affected by stress, anxiety, or depression loses about sixteen workdays a year.

The Northwestern National Life Insurance Company did a survey in 1991 of six hundred U.S. workers. Thirty-four percent of the respondents considered quitting in 1990 due to job stress. If you quit your job because of stress, you are one of many who made that decision (about one in every six workers joined you).

Here Today, Gone Tomorrow

As a result of two hectic decades of takeovers, bankruptcies, restructurings, downsizings, and relocations in business (and layoffs in government, too), workforce morale is about as strong and high as a dead bird. This is not a good way for us to be preparing for the global competitiveness of the twenty-first century. The morning sound of the alarm clocks is met by increasing numbers of workers with a sense of dread and foreboding. While we watch the front lines and we see many of our comrades being picked off by rapid-fire pink slips, the best most of us can hope to do nowadays is work quietly in the trenches, trying not to be noticed.

With financial analysts at the helm, their spreadsheets wallpapering the office, it often looks in the executive suite as though the best—maybe the only—way to cut costs is by cutting staff. But beyond the initial boost to the bottom line, that may actually be a strategy that boomerangs. Because what results is increased turnover, reduced productivity, more absenteeism and illness due to the stress felt by those who are left on the payroll. Part of this stress is guilt. Part of it is dread about whose turn is next. Part of it is what is often perceived as the inept and unfair manner in which most downsizing is done. Part of it is cynicism about and disbelief in management communications, which leads to the rampant growth of the "grapevine," which ultimately strangles all official memos from the executive office.

What's more, employees who have been laid off already know their fate, and their employers have usually handed them out-placement and/or severance packages. Those of you who remain at work are usually given nothing in the way of support or sympathy except for added work loads, not to mention the fact that you may no longer experience any comfort in a sense of loyalty, career goals, and future security. Whether working for yourself or someone else, the future of your job appears far less certain, and if you hate what you do, you feel trapped anyway, because the job market is the pits. So, while the paycheck may keep arriving for now, it's very likely that it will be accompanied by the ulcers, headaches, and other assorted effects of stress.

Robert Karasek and Tores Theorell, in their book *Healthy Work: Stress, Productivity and the Reconstruction of Working Life*, report that people under thirty in the United States are more than three times as likely to suffer from depression as their grandparents. Global interdependence and production overcapacity, leading to job insecurity, is cited as a prime source of the stress.

But, clearly, it's not the only source.

"Because I'm the Boss!"

In the study that was done by Northwestern National Life Insurance, the researchers concluded that while workers often blame overwork for their stress, the real cause may be lack of control over one's job.

The increase in both middle management and mid-level professionals, and the size of the Baby Boom, have come together to create an important bulge in the workforce. This bulge is made up of people who believe they have the right to make decisions about their work and possess the training and education to exercise good judgment. Yet, their perception is that they have less freedom to use these. As a result, those in this bulge—Boomers, middle managers, employed professionals—have become very vulnerable to stress and stress-related illnesses.

Sometimes the restricted use of your own judgment is a result of your having a "dysfunctional," or emotionally scarred, supervisor or boss. The Bureau of National Affairs conducted a fascinating and much discussed study in 1990 that concluded that as many as one in four bosses manage their offices the way their parents ran their homes—in ways that were demeaning and insulting, with a lot of abusive behavior that greatly depressed morale and undermined productivity. The report stated that employees who are exposed to and abused by these managers have more health problems and difficulty concentrating. These kinds of bosses cause their offices to suffer from higher health care costs, absenteeism, turnover, and poor performance. The question is, why is it that a dysfunctional supervisor is not treated in the same manner as one who suffers from substance abuse? The effects on people and work can be identical.

A second source of lack of control over your job could be the nature of the job you're in.

Robert Karasek, an industrial engineer at the University of Southern California, has defined four categories of jobs, depending on how much stress they cause. He based his model on how demanding the task is and how much control the worker has. "Demand" relates to how hectic and chaotic the workplace is, how many requests are made of the employee in a short time, how pressured the job is and how often and difficult the deadlines are. "Control" implies the ability to make decisions about your own job, about the jobs of others, about your work load, work method, and even the way your resources (money, time, help) are gotten and used.

Those people who experience the most stress are those who are in high-demand, low-control jobs. They face constant pressure to report, produce, serve, finish, respond, etc., while at the same time their opportunities to decide just how best to do those things are limited.

If this is your situation, you are drinking a dangerous cocktail. You should be regularly monitoring your blood pressure to make sure you are in good health and able to withstand the physical punishment as a result of the emotional anguish you are likely to be experiencing. According to this theory, you are at higher risk than average for stress and depression.

Karasek's model is now being supported by research and analysis from a number of others. Psychologist Mariquita Mullan, a former public health officer at the National Institute for Occupational Safety and Health, observed this when she was studying workplace stress in fire fighters in New York City. Those whose tasks gave them more individual control were reporting less stress than those who took orders as part of a team. Jeffrey Johnson, associate professor at Johns Hopkins University, concurs. He feels there are people who are not getting much satisfaction from their work, and at the same time they are being asked to do too much. Also at Johns Hopkins, epidemiologist William Eaton and others published a study of nearly twelve thousand workers in the *Journal of Occupational Medicine* in 1990. The link between occupation, stress, and depression was very strong, and correlated highly with how much autonomy (self-control) members of an occupation felt they had.

A surprising result was that lawyers were 3.6 times more likely than average workers to become clinically depressed, placing them at the top of over one hundred occupational groups.* Secretaries and some educators (particularly school counselors and special education teachers) were also two to three times more likely to experience depression than the typical worker.

Eaton and others now believe that wherever there's one or more bosses who frequently say, "Stop that and do this," or "Do it my way," or "Don't think, just do it," heightened levels of stress and depression are almost inevitable. The ten worst professions, according to the Johns Hopkins study, are: data-entry keyers; computer equipment operators; special education teachers and school counselors; typists; lawyers; health aides; production inspectors; waiters and waitresses; food preparation workers; and sales workers.

It is difficult to know whether highly stressed, depressive people seek these particular jobs out, or whether the jobs cause the symptoms. In any event, the National Institute of Mental Health, reporting on a massive study, found that 4 percent of American workers suffer job-related depression, with symptoms that include prolonged sadness, sleep and appetite disorders, difficulty concentrating, and even thoughts of suicide. Dr. Peter Schnall of Cornell University is among those whose research has concluded that chronic job stress as a result of low control over work can elevate pulse, pressure, fatigue, levels of substance abuse, absenteeism, heart disease, asthma, headaches, ulcers, and possibly even arthritis. Just something to consider when we think about what we want to be when we grow up.

*Psychologists Everett Moitoza (New Hampshire) and Andrew Benjamin (Washington State), who have had extensive experience treating lawyers, attributed the startlingly high rate of depression to growing pressures: the need to increase billable hours; the adversarial nature of the profession, which caused many interpersonal problems within and outside their practice; the inability to accept the kinds of cases they were interested in and went into law to pursue; the constant and often tedious attention to process as opposed to challenge and ideology; overwhelming time commitment that leads to enormous burdens on any family or social life they may desire. One American Bar Association (ABA) official wrote of the "deterioration of the lawyer workplace."

Is it any wonder that our grandfathers were never so much impressed with our diplomas, but rather shook a finger at us and said, "Be your own boss!"

No Time to Smell the Flowers

Whether we suffer stress as a result of job insecurity or high demand/low control situations, one thing is for sure. In these situations and others, time seems to have shrunk along with the waistlines of our clothing. And while we want to free some time to exercise and lose the weight, the hands of the clock keep pointing out that there's no time to spare. Like Alice through the looking glass, we run harder and harder, faster and faster, just to stay in place. January passes, spring passes, the year passes, our youth passes, our children are grown-up strangers or never born, in all cases often because we are too busy working to notice.

As organizations seek to operate "lean and mean," it is the "mean" that seems increasingly to dominate. According to *Fortune* magazine, the number of managers who say they have too much to do has jumped from 34 percent to 46 percent in the past five years. Organizations often reduce the number of people employed, but not the work. Karasek and Theorell state in their book that working hours increased from 40.6 hours per week in 1973 to 48.8 hours per week in 1988, while real income dropped in that same period.

Harvard economist Juliet Schor, author of *The Overworked American: The Unexpected Decline of Leisure*, also contradicts the belief that today's workers have more leisure time than ever. Her research leads her to conclude that the average employed American will today work the equivalent of one month more per year than the typical worker in 1970. She believes U.S. employees put in more time working than those in any other industrialized country, except Japan, where dissatisfaction rates are rising sharply. And if the projections of this trend continue unchallenged, Schor predicts that in the year 2010, we will be at our work sixty hours a week, fifty weeks a year. Schor makes a strong case for the need to pay serious attention to the degradation of life this is leading to. "Those who call for America to replicate the Japanese work culture have forgotten that the point of economic success is to make possible a good life. To impair the quality of life in the name of economic success is foolhardy."

This is leading to high, and growing, rates of stress and stress-related diseases, especially for women who find that the competing pressures to have a family, care for a family, and care for themselves are all demanding time and attention that seem difficult to spare—particularly when faced with a "biological clock."

The ultimate health problem here, though, is "being worked to death." Does that sound like just a cliché? Well, in Japan, *karoshi*, a term meaning sudden, premature death from overwork, has become a legal and recognizable issue. Fukoku Life Insurance Company and the Japanese government, in studies of Japanese workers, found that almost 43 percent of salaried employees in Tokyo feared their jobs would lead to *karoshi*, and almost three-fourths felt stressed. Over 40 percent felt constant fatigue, and more than one in five often wanted to call in sick. More than 85 percent of Japanese workers just wanted more sleep, and by a two to one margin, they preferred more free time to more pay. The Karoshi Emergency Call, a Tokyo hotline begun in 1988, has received almost two thousand calls, half of them involving death. But the Labor Ministry resists recognizing *karoshi* as a work-related issue, because it says the relationship between circulatory diseases and fatigue has not been proved.

Not everyone agrees that U.S. workers actually *work* more hours than ever before. Some researchers, such as John Robinson, professor of sociology at the University of Maryland, feel Americans have more free time, but it only feels like less.

He says we spend *fewer* hours working today than twenty-five years ago, and we're doing significantly less housework as well. But he admits that a third of all U.S. workers between the ages of eighteen and sixty-four report they always feel rushed to do the things they have to do, up from 28 percent in 1975 and 25 percent in 1965. Those aged thirty-five to fifty-four are always more likely to feel this way, and the growth of that age group may be why there seems to be an increase in this type of stress in the general population. And, as in the other studies, women are shown to feel more squeezed for time than men. But outside of the dual family/work role many women now face, Robinson thinks the perception of time squeeze is not a result of paid work but of growing organizational membership, athletic activity, TV viewing, and interpersonal time.

Researchers on both sides of the controversy—more work time, less work time—still point to a heightened sense of time-related stress in the population. Michael Ventura wrote several years ago in the *Whole Earth Review* that the boundaries "between intimate time and business time; between home and work; between night and day; between individual and corporate; between private space and public space" have all become blurred in the last few years. Time has become amorphous. We are all in the throes of adjusting to "timelessness"—a sort of chaos in which we are being asked to define for ourselves the nature, speed, and boundaries of time.*

Not an easy task for mere mortals trying to adapt their biology to the modern world of work.

*We need to give more thought to how the computer has affected all this. Clearly, computers and such computer-related concepts as "real time," "virtual reality," and "on-line" have profoundly distorted our customary time-space parameters. Time has speeded up, and distance has practically disappeared. What does all this do to, for example, our biorhythms? We need to know.

Juggling Job and Family

We already know that there is a great deal of stress experienced by women who must juggle the competing demands of work and family. Problems with children, spouses, and parents are likely to get to the pulse, pressure, and gastric systems of women on a regular basis. Many are capable of withstanding the physical assault. Others eventually succumb to the pressure, becoming physically ill, emotionally badgered, or both.

Now we are learning that this is a man's as well as a woman's issue. When women need to work outside the home, men are finding they are under increased pressure to spend more time with their children and to help with the household chores. And because most work environments are even *less* prepared to allow male workers to attend to family life, the stress felt by men may actually start to become greater than that felt by women.

A 1988 Boston University study of sixteen hundred employees at a utility and a technology company in the Northeast found that only about one worker in ten was part of the "traditional" workforce—defined as married males with nonworking wives. And as many fathers as mothers (36 percent versus 37 percent) reported feeling a lot of stress when trying to balance their work and family lives. In fact, men experienced *more* stress than women when child care duties forced them to miss work. Under such situations, men are more likely to report health problems, such as headaches, bad moods, and unhappiness with their jobs or families.

Study after study is now showing that, in selected ways, men are suffering as much as, if not more than, women in the work/family arena. In order to reduce stress, spouse and children are increasingly becoming factors for men when they set work goals, reject transfers, turn down promotions, refuse overtime, and reduce travel.

Who Helps the Helper?

"Burnout" is the term used when people feel downright fatigued from the work they do—tired of the job, tired of the hours, tired of the pressures. At its extreme, burnout leads to a kind of alienation—removal from caring about the work, just wanting to get away and go to sleep. Yet burnout also causes insomnia and problems with home life.

Many who work in the field of providing aid, comfort, and other human services find that burnout is especially painful. Social workers, psychologists, nurses, doctors, teachers—these are all people who are asked over and over again to absorb the ills and suffering of other people as well as their own. Whether dealing with drug addicts or their children, battered women or their imprisoned husbands, dying young men or aging old women, those in care-giving jobs are at higher risk of getting to a point where they start to experience constant irritability, personal depression, muscle aches and headaches, and a sense of despair.

Michael Groh, a management consultant with the Amherst H. Wilder Foundation, has developed a seminar on burnout, especially designed for human services workers. He stresses diet, fitness, and relaxation, as well as modification of the workplace to reduce noise and interruptions. He's also big on time management and time off, all central to the ability to ward off burnout.

If the Shoe Fits . . .

Groh's emphasis on redesigning the workplace is important to note, because we've already seen in chapter 7 that ergonomics plays a large role in our physical well-being.

Because more and more of us work with computers, we want to be sure to clue you in on the fact that, especially if you are a woman, it is critical that you "try on" your equipment before buying, choosing, or settling into it. Many workstations are designed not only for male proportions, but also for male sensory faculties. And there *is* a definite link to stress.

Caroline Dow and Douglas Covert of the University of Evansville in Indiana found that women who use computer monitors that produce 16-kilohertz background noise quickly experience stress, which leads to inefficiency and lower performance scores. Not only do women hear sounds at higher frequency than men, but they're most sensitive at the high estrogen points of their menstrual cycles. In some, this heightened stress leads to headaches, for others to increased blood pressure, and still others suffer from muscle tension, irritability, and combativeness. Not all monitors are a problem. So, just like buying a pair of shoes, it is important that you try on your equipment. It's best to use it for at least thirty minutes to an hour to test how it "wears" on your mind/body system.

The Race for Space

A large body of research and many books and movies depict the crowding and congestion of modern times and the increased stress that results. Like rats, and other creatures, we humans react badly to overcrowding. When put in a box and allowed to overpopulate, rats wind up biting each other, savaging their young, and becoming alienated and vicious. Welcome to the freeway at rush hour. Or the subway. Or the city bus.

Do you remember that term "rush hour"? Does anything "rush" on rush hour? And does it only last for an hour? In 1950, there were about 50 million U.S. commuters. Today, there are over 100 million. According to one population study, in the next thirty years, the United States will grow by almost 90 million people and over 40 million cars. Congestion will increase by the year 2010 to five times the present amount on the nation's freeways, and triple what it is on other roads. In California now, the average travel speed is 33 miles per hour. It is expected to drop to 15 miles per hour by the dawn of the twenty-first century.

And the suburbs are taking the brunt of much of the congestion. Today's worst gridlock and stoppages often take place far from the urban cores, in outlying suburban areas that have transformed themselves into business hubs.

More and more people arise before first light, to fight their way into the office, only to anticipate the painfully crowded and slow trip home. A neon sign that flashes the word STRESS in bright colors should probably be hung in the skies around our commuting paths, to remind us of the toll that is being taken on our minds and bodies every day, day in and day out. It is no wonder that more and more, we are accommodating to the idea of working closer to home—of using technology, decentralized office locations, job-sharing routines, and flexible weeks in order to cut down on the time, energy, and stress involved in just getting to and from work.

Private Space

We are not always conscious of our basic need for private space. Nor do we know all the consequences to our body and psyche when we are deprived of it. But we do know that there *are* impacts. The office washroom is a case in point. Observers have noticed that when two male acquaintances are standing side by side at urinals, they may engage in pleasant, ordinary conversation, but they will never look at each other. And if for any reason one needs to turn to look at the other, the gaze is limited to above the belt. If the two men using the adjacent urinals are strangers, however, they will behave as though standing next to them there is no one at all, only a void.

In his wonderful book, *The Science of Everyday Life*, Jay

Ingram reports on a 1976 study of how stress comes into play in the washroom. Research had already confirmed that fear and anxiety could produce an inability to urinate. A person who needs to void is faced with a biological problem while he or she is fearful—the bladder muscles contract, forcing urine into the urethra, but the sphincter muscles (which must relax to let the urine out) are paralyzed by fear. At a midwestern U.S. university, a washroom was used as a stress laboratory. It contained two toilet stalls and three urinals. Under the conditions of the experiment, the unknowing subjects who entered the washroom could have found themselves in any one of these situations: they were alone, and the other two urinals were supposedly not working ("Don't Use" signs were on them); or there was one other man (working with the experimenter) separated from the subject by a "Don't Use" urinal; or the other man was standing at the urinal right next to the subject, with the one at the other end bearing a "Don't Use" sign. Using an elaborate setup with a periscope (listening didn't work with these urinals; they were too quiet), a researcher in the stall worked with two stopwatches, one to time how long it took for each research subject to begin to urinate, and one to clock the length of time of the flow.

The results bore out the stressful impact on the body of invasion of our most private space. When the subjects had the washroom to themselves, the delay in the onset of urination averaged 4.9 seconds. The presence of a stranger one urinal removed had the effect of delaying onset to 6.2 seconds. And if the subject and the collaborator were standing side by side, the delay grew to 8.4 seconds.

Although there is some debate about the ethics of this experimenter's methods, the results are not at all surprising to any of us. In mathematics, the results are considered statistically significant. In life, the results are called commonplace. We all know how important private space is to us, in some circumstances more than others, and when the workspace or the washroom or the freeway gets to where we feel the stress, our body rebels.

US Air ran an ad in 1992 in which they cited a number of studies that point to the importance of personal space. In one, 120 undergraduates in a large midwestern university were randomly assigned to either a crowded or uncrowded seating arrangement when taking an exam. The subjects with less personal space experienced higher anxiety and scored lower on the test than those in the other group. This is something you may want to consider when selecting a job or choosing your workspace.

Cape Fear

Another source of stress in the workplace revolves around our growing concern for personal safety and security. For example, with each day come new reports on weaponry, particularly concerns about who is buying it, and whether or why it might be brought into the workplace.

The executive protection business is estimated to be over $1 billion a year in the United States. This includes kidnap and ransom insurance, bodyguards, and armored cars. In 1990, *Fortune* reported that while terrorists were a real problem, disgruntled employees were the greater threat. One clinical psychologist knew of fifteen incidents in 1989 in which chief executives were lectured to at gunpoint in their offices.

With rising violence rates and shrinking police budgets, private security forces are growing. But there are some who believe that many of the security guards employed today in the United States are unqualified, dishonest, unreliable, and violent.

Does all this make you feel safe and secure? Lower your stress levels? The best advice here is to use your leisure time to see comedy and family movies. Injections of lighthearted unreality may be just what the doctor orders. And if you have reason to believe that your life or office truly is or could be in danger, don't take security lightly. If you hire it or install it, do it well and do it right. In the end, this is no laughing matter.

It Takes All Kinds

For several decades, analysts have been trying to come to grips with just what the issue of stress is all about. We never really paid much attention to it before the mid-seventies. Certainly, it existed. It just didn't seem to have as many interested and vocal people suffering from it, studying it, making money on it, writing and lecturing about it, designing programs around it, or putting in insurance claims for it.

At one point in the eighties we heard talk of two kinds of stress—positive (or eustress) and negative. Positive stress is said to beneficially challenge the system's energy, so it can produce creativity and work. People who respond positively to deadlines—reporters and advertising copywriters—are thought to represent eustress. Negative stress, on the other hand, drains enthusiasm and innovation, leaving slackened output and weakened health.

A spate of studies also began to sort through personality types, to define people on the basis of their symptoms and their tolerance for stress. For example, the National Headache Foundation in Chicago has linked personality traits with headache type. *Migraine* sufferers, they say, tend to be neat, orderly, and creative people who demand a lot of themselves. Their pain is not always one-sided or limited to the temple. Those who suffer from *cluster* headaches, in which the pain is almost always one-sided, are usually present or former smokers with aggressive personalities. They often experience feelings of restlessness. And *tension* headache sufferers are the ones with poor coping and adaptive skills for stress. They tend to be irritable, easily agitated, and have a tendency toward depression. They can get a headache just from thinking about being under stress. Their headaches tend to recur in the same place.

Hostility, too, has been linked to physical illness. While the components of hostility—aggression, noncooperation, a hot temper—might serve some people well in the competitive world of business, it is a form of stress that could take years off your life. Doctors at the Duke University Medical Center have designed a psychological test that measures hostility, high scores on which have also been predictive of heart disease. In fact, high hostility scores boosted the risk of death from all causes.

Yet some scientists believe there is still no conclusive link between psychological factors and immune function. Psychoneuroimmunology, the science that studies this, has strong proponents and opponents. Many researchers believe there has not yet been any conclusive evidence linking stress to the appearance of any specific disease.

But other research continues to point to the existence of such links. For example, studies have now demonstrated that compared with lowest-stress groups, subjects with high psychological stress had double the incidence of getting a common cold and five times the incidence of infection with a cold virus.

The existence of positive stress aside, it seems that stress is an invisible biological weapon, firing on us and at us for so many reasons so much of the time, causing serious wounds. As we turn our attention to stress in the office, there are a number of things we can do to diminish the negative effects.

Healthy Companies

It is interesting to study just why it is that some organizations have significantly lower health care costs than the majority, who are crippled by the escalating costs of the past few decades. Morty Lefkoe, a Connecticut management consultant, writing in *Across the Board* magazine, notes that companies such as W. L. Gore & Associates, North American Tool & Die, Inc., Leaf, Inc., Johnsonville Foods, Inc., and Quad/Graphics are examples of those that pay attention to developing a corporate culture that reduces stress. At Gore, everyone's title is "associate," and work is organized by self-determined commitments. Trust and open communication are emphasized. North American Tool & Die emphasizes safety and security, especially of work processes, looking to enhance worker "happiness." At Johnsonville Foods, worker autonomy and cooperation are the cornerstone of all operations. At Quad/Graphics there is no such thing as "Because I'm the boss, that's why!"

Study after study has shown that employers who reduce the aspects of stress we discussed in this chapter not only get better work from their employees, they also save a great deal in health care costs.*

But Veronica Goff, manager of mental health promotion at the Washington Business Group on Health, laments that, despite all the research that links mind-body problems in the office to health costs and productivity, there is still hardly any attention being paid by employers to your emotions, attitudes, and psychological environment and well-being. "Caring" is still not a part of management training.

So, for most of you, stress reduction is your responsibility for now.

Some Stress-Reducing Tips

- Try out several *relaxation techniques* to see which might work for you and under what circumstances. Some people

*For more on this, read Robert Rosen's *The Healthy Company*. Rosen is president of Healthy Companies, a nonprofit group funded by the MacArthur Foundation.

find yoga, or transcendental meditation, or even religious or spiritual meditation, to be extremely helpful. (Interestingly, Silver Cross—Holland's largest private health insurer—offers a premium discount of 30 percent to individuals who practice transcendental meditation.) Some of these relaxation techniques are extensive, requiring hours of stillness or concentration each week. Others involve only short and periodic breaks from a busy work schedule while at the office—e.g., stretching and bending for two minutes each hour, or neck and shoulder rolls while talking on the telephone. We won't recommend any particular one (there are literally dozens of books you can select exercises from, and probably almost all will be beneficial).

One thing we do believe is particularly good for reducing stress is deep-breathing breaks. After a particularly stressful incident, hour, morning, or day, anyone who doesn't have a serious heart problem could make a conscious effort to calm down by doing this: sit up straight in your chair, with your feet together, flat on the floor, and your hands placed flat in your lap. Make sure your back and shoulders are straight (not arched), and close your eyes. Try to imagine that your hands and feet are tingling. Then begin to breathe *so you can hear it.* Listen as you draw the air in, first as if into your stomach (and expand your stomach as you do so), and then into your lungs (and expand your chest as you do so). Breathe in slowly to the count of ten ("one-alligator—two-alligator—three-alligator," etc.). Breathe out, first by drawing in the lungs and then collapsing the stomach, to the count of thirteen. You will find that it is the breathing *out* rather than *in* that not only challenges the stress in the system, but ultimately relaxes it. Repeat this ten times, slowly open your eyes, then shake your hands for about fifteen seconds.

You will find that some version of this breathing technique appears in almost all instructions for self-hypnosis and asthma control as well as biofeedback training. Biofeedback is another important relaxation technique that utilizes small machines to monitor physical stress and, through audio or visual signals, helps you reduce that stress. These are devices that provide you with clues as to the tension in your muscles, or your nerves, or even your brain waves. And it is amazing how, when you are hooked up to these devices, thoughts of rain forests calm you, but thoughts of your boss, your job, or your co-workers send the signals through the roof. Biofeedback has helped many people learn to calm their body's stressful reactions to the things that provoke them. If you believe you are suffering stress that is troublesome to you, we suggest you check this out.

- *Distract your mind by refocusing your senses.* As we pointed out in chapter 1, certain scents such as muguet (lily of the valley), or roses, or lavender, or apple spice may have a calming effect on your system. You may want to experiment with these yourself or consult an aromatherapist. (Watch out for allergies; some people are susceptible.) Buy a book of puzzles that challenge you but that you enjoy—crossword, logic, seek-a-word—and tackle one or two. Your eyes and thoughts will be focused on this task. Or play Gameboy. Listen to a tape that you especially love. (Did you notice that during the 1992 Summer Olympics the athletes, particularly the divers, had earphones on between rounds?)
- *Drop your jaw!* Many researchers who study stress have found that the jaw is often the first indication of trouble. When stressed, we tend to tighten our jaw and, in the process, grind our teeth (dentists know all too well the toll that stress takes on teeth and gums). Tightened jaws also help bring on headaches. So, whether or not you are conscious of stress, do a periodic "jaw check"—once each hour, relax your jaw and, if you can, the rest of your facial muscles. You'll be surprised at the instant impact this seems to have on your feelings of tenseness.
- *Time management* can be a significant source of stress reduc-

tion. At some point, set aside some hours to review thoughtfully and honestly how you use your time. If you cannot control the way your time is used while at the office, then reprogram and reevaluate your nonoffice time. Who is taking up too much time unnecessarily? Put a stop to that. Who is not getting enough of your time? Plan to increase it. How? Figure out what you are doing that you really don't need to do. Make trade-offs. You'll fret at first, but you'll find over time that you really won't miss what you've let go, providing what you've replaced it with is really meaningful.

Also—make lists! Whatever it is, write it down! A lot of stress that seems to come from not having enough time is really your mind reviewing over and over again a series of your lists—what you have to do, where you have to go, what you have to buy, what you have to prepare, whom you have to call, what you need to say, what you have to pay, etc.—which, once written down, can be dumped from your mind! Then, when the temptation comes to do the mental Olympics of reviewing the mass of activity, you can relax and remind yourself that it's written down, and as you attend to each, you can cross it off or add another.

- *Laugh.* A number of researchers are hard at work exploring the impacts that humor has on stress. There is increasing evidence that, by engaging much of the brain in an enjoyable diversion, humor does indeed improve physical health. Lee Berk, a psychoneuroimmunologist working with associates at the Loma Linda University Schools of Medicine and Public Health in California, has been studying neurochemical changes caused by laughter. Comedy causes research subjects to have lowered levels of the neurotransmitter epinephrine and the stress hormone cortisol. Lowered epinephrine can reduce high blood pressure and diminish other cardiovascular problems. Reduced cortisol permits greater production of beneficial white blood cells, boosting immunity. So, it makes great sense to surround yourself with people who have a wonderful sense of humor. If you can't seem to touch base with these people often enough, read more of Woody Allen, Wendy Wasserstein, and whatever works of humor you can get your hands on. Also, find a video of *It's a Mad, Mad, Mad, Mad World*. It's a great movie, bound to make you laugh. And if your cable station doesn't offer the Comedy Channel, get up a petition!

- *Get a life!* Perhaps one of the best ways of managing office stress is starting to attend to other things in life that matter outside the office. Make conscious efforts to refocus on existing relationships and build new ones. Pump romance, love, fun, camaraderie, learning, intimacy, caring, and community back into your life. Make time for friends, and make some new friends. Take a class or two on something(s) you've always wanted to know about. Fix the tension that spilled over from work to home with humor, surprises, short trips to the city, or long hikes in the country. Take a new look at your core values. Are you cut out to give 150 percent to the office with nothing left over for you? And by doing that, do you get so stressed out that, in reality, the office is only getting 65 percent and there's still nothing left for you?

As we enter the twenty-first century, the work we have traditionally done will require a smaller percentage of the growing population. More and more of us will have to reinvent our jobs, our careers, and the balance of our lives. If we are to reduce the physical illnesses that are bound to result from the enormous stress we will experience, we will have to truly learn what it means to get a life!

11 *Substance Abuse: Workplace Nightmare*

If you have been working for some time, you have probably had the experience of working with or for a substance abuser. In most cases (and the most obvious cases), that person was an alcoholic. Probably he or she had a liquid lunch and for the rest of the day was either useless or painfully disruptive.

Until recently, particularly if the abuser was in management, the problem was ignored or somehow gotten around. It was seen as the problem of the individual and not as a problem of the organization.

No more. The situation began changing in the eighties, in large measure because employers became aware of the costs of tolerating substance abuse. A study reported on by the Conference Board in 1989 concluded that American business was losing an amount equivalent to 3 percent of payroll each year because of drug and alcohol abuse by employees. A federal government report at around the same time estimated that substance abuse was costing the economy almost $200 billion a year. General Motors said a few years ago that drug abuse cost it more than $1 billion a year—largely for treatment and absenteeism.

The costs we can measure come from increases in health insurance prices, use of health care services, absenteeism, reduced productivity, and theft and embezzlement. There are hidden costs as well: disruptions and reduced morale in the workforce, impaired relations with co-workers, customers, and suppliers, increased crime—just to name a few.

The Extent of the Problem

In spite of all our efforts, in spite of government programs and public education efforts, substance abuse remains a huge problem. A 1988 report by the New York State Division of Substance Abuse Services stated that almost one-third of adults had used an illicit drug at least once in their lives. And while you may think this is a lower-class problem, the highest rate of abuse was in households with incomes of over $100,000 per year.

Americans spend well over $100 billion a year for illegal drugs. The most widely used illicit drug in the workplace is cocaine. According to knowledgeable observers, cocaine is a very democratic drug; its users range from janitors to CEOs, although it is particularly prevalent among younger people. One of the insidious aspects of cocaine, say many experts, is that addicts tend to abuse other substances as well. They can end up with multiple addictions, involving alcohol, heroin, and other drugs.

Crack cocaine destroys people quickly. People may drink for many years before they become alcoholics. Crack smokers can become addicts in a few weeks or sooner.

There is increasing documentation showing that cocaine causes brain damage. Neurological studies show that habitual cocaine use can result in cerebral atrophy. In a Minnesota study of cocaine addicts, for example, more than half the addicts had indications of brain atrophy.

Other drugs are also dangerous. Ecstasy, a popular "designer" drug, is known to damage brain cells. The most widely used and abused drug is alcohol. It has been estimated that costs to society from alcohol abuse are three times those due to use of illicit drugs. That includes costs of reduced productivity, motor vehicle accidents, and alcohol-related diseases such as cancer and cirrhosis of the liver.

Disease or Life-style?

There is currently a fierce controversy over alcoholism: is it, as most people claim, a disease, or is it an indulgence? The answer, if we ever arrive at it, has enormous social and economic implications for society.

If alcoholism is a disease, are alcoholics helpless victims of it who cannot be held accountable for their actions? Indeed, their helplessness requires that they receive assistance—from you, from others, from employers, from government. Furthermore, addiction can be classified as a disability. It is conceivable, therefore, that under the provisions of the Americans with Disabilities Act (ADA), alcoholics cannot be discriminated against in hiring, firing, and promotion decisions. A physiological flaw, such as addiction, cannot be held against an individual. As one writer said, "The more biological a given behavior is, the less control the behaver has over it."

This, of course, raises questions about helplessness and individual responsibility as they relate to a host of other human traits and activities. If we accept that alcoholism comes originally from the genetic makeup of the individual, can we say the same thing about any behavior that appears compulsive or obsessive? Are all such behaviors diseases?

In recent years, the courts have tended to broaden liability of parties who permit or encourage use of alcohol that results in death, injury, or property damage. Bars and bartenders who serve liquor to intoxicated patrons, for example. And increasingly, employers are being held accountable for the consequences of behavior by their employees. If you got drunk at a company party and had an accident on the way home, your employer might be liable.

The more we learn about our genetic makeup, the more our genes appear to be responsible for our behavioral as well as physical manifestations. Are we, as some scientists would have it, programmed machines that can no more be held responsible for malfunctions than computers or automobiles?

Most mental health practitioners today favor the addiction-as-disease approach. This can be seen, for example, in an article in the (now extinct) publication *Behavior Today*. The article pointed out that all who were involved with or related to an alcoholic—family, friends, co-workers, employers—might be "enablers," encouraging the alcoholic by helping to protect him from the consequences of his behavior.

Another addiction drawing considerable attention these days is cigarette smoking. (It must be admitted, though, that this one is treated with less tolerance now than alcoholism: the customary approach for tobacco abuse is prohibition.) Cigarette smoking, like drinking alcoholic beverages, results in horrendous costs. The Johnson & Johnson company estimates that up to one-quarter of health care costs borne by business derive from what they call the "unhealthy life-style conditions" of employees. Perhaps the two biggest contributors to negative life-style conditions are alcohol and cigarettes. With business costs for health increasing at the rate of almost 10 percent a year, the situation is rapidly becoming intolerable.

So businesses are acting to hold down or eliminate costs. They are more and more looking at drinkers and smokers as high-risk people and therefore undesirable. In some cases, they are making smokers pay more for health insurance than nonsmokers. Last year, for example, electronics giant Texas Instruments instituted a $10 per month extra charge on health insurance for employers who smoke—including away from work. The same charge applies to any dependents who smoke.

Other companies are using the carrot rather than the stick. They encourage and even pay for treatment and counseling programs for drinkers and smokers.

There is reason for optimism at least with regard to cigarette smoking. The latest report from the Centers for Disease Control shows that in 1990 smoking was at its lowest level since the centers began collecting statistics on smoking. Twenty-five percent of adult Americans smoked cigarettes in 1990, continuing the steady decline that began in 1965. A question: Does the significant success enjoyed by the anti-smoking people indicate that their relatively harsh approach to addiction can work for other abused substances?

Alcohol, tobacco, and recreational drugs such as heroin and cocaine are not the only addiction problems in the workplace. Steroid abuse is becoming a problem. By the late 1980s, it was estimated that one in fifteen male high school seniors used anabolic steroids. Permanent physical and mental damage can result. All-pro defensive lineman Lyle Alzado blamed his prolonged and extensive use of steroids for the brain cancer that killed him at age forty-three.

Aside from substances, there are some behaviors that appear to be addictive. One is gambling. Another is high-risk sports, such as sky diving and bungee jumping, whose practitioners are called "stimulus addicts" by psychologists.

In 1988, a federal court ruled that an FBI agent who was fired for gambling away $2,000 in government funds was not a criminal but was instead handicapped by his "pathological gambling."

Thrill seekers, also known as Type Ts, are, according to psychologist Frank Farley of the University of Wisconsin, responding to a chemical condition in their brains not unlike that of drug addicts. People in both categories require greater stimulation than others; one group gets it from thrills, and the other from drugs. We could speculate that Type Ts may be likely to get involved with white collar crime and shoplifting.

Although the focus of substance use is on the abusers—the addicts—some observers say that casual or occasional use, unaccompanied by addiction, is pervasive in our society. Apparently, there are many more of us who can use alcohol, tobacco, and illicit drugs without becoming addicted than there are people who get hooked.

One scientist calls the casual users "chippers." They display no signs of dependency, and they are able to stop using whatever substance they indulge in without the discomfort of withdrawal symptoms. They may be psychologically different from addicts: less subject to stress, more able to be in control. This may be accompanied, or even caused, by biological differences.

And it is these biological differences that are so intriguing and that seem to be leading us now to new perceptions and insights into human behavior. Some scientists now equate the need for mood-altering substances with the three primary driving needs: hunger, thirst, and sex. If this new approach is valid, it follows logically that we cannot stamp out addiction by law and law enforcement. (Nor, it should be pointed out, will legalizing drugs reduce drug abuse if the biological drive theory is true. It is more likely instead that drug use will increase, perhaps even approaching the extent of alcohol and tobacco use.)

Some people are concerned that the existence of a powerful biological need for mood alteration will make drug control efforts futile. They propose that we concentrate our efforts on creating alternatives. Some are exploring the natural "drugs" secreted by our bodies that delight and excite our brains, such as dopamine and endorphin. Others are seeking to develop harmless drugs—"utopiants"—that will provide pleasure without damage. It may even be possible, they say, to develop such safe drugs with built-in chemical agents that will prevent abuse and the extremes of addiction. (It might be instructive to recall that heroin was originally developed as a "safe" alternative to morphine.)

The Propranolol Dilemma

Another dilemma involving the use of mood-enhancing substances is the "abuse" of legal drugs. Propranolol, a beta-blocking prescription drug, is used in the treatment of cardiovascular conditions: angina, cardiac arrhythmia, and hypertension. It is also used to treat migraine headaches and anxiety resulting from stress. Propranolol blocks the adrenaline-like chemicals that the body overproduces when under stress. In effect, then, it can diminish the physical symptoms of stress—rapid heartbeat, sweating, dry mouth, etc.

The Hastings Center is an institution that examines and reports on the ethical and moral aspects of health issues. In a recent issue of its magazine, *Hastings Center Report,* bioethicist Jacquelyn Slomka of the Cleveland Clinic Foundation addressed some dilemmas involving use of propranolol, particularly in light of its apparent lack of adverse side effects.

It seems that performers have discovered that propranolol, by reducing the physical symptoms associated with performance anxiety, is in effect a performance-enhancing substance. Propranolol can dispel nervousness without in any way affecting mental alertness and creativity.

A survey of musicians done in 1987 showed that almost 30 percent of them used a beta-blocker (mostly propranolol) before auditions and performances. Doctors, who are obviously knowledgeable about propranolol, are also users, although the extent of their use is unknown.

Is it morally acceptable to use performance enhancers? We answer no in the case of steroid use by athletes, but most people say that's because steroids can be harmful. Yet it's not that simple. Many people also say that steroid use is also morally wrong because it is an unnatural affecter of performance. Even some users seem to experience a form of self-loathing, as if they are repelled by their reliance on a "drug."

As Slomka asks, is the use of drugs to improve performance a violation of the American value of fairness and a belief that hard work is the way to earn rewards? Is performance anxiety a disease? If it is, is the regular use of propranolol a legitimate treatment? Or is it merely an ordinary biological response to extraordinary circumstances? If the answer to this last question is yes, does the use of a drug to treat it eventually result in a loss of our ability to marshal our own resources to deal with even the small traumas of everyday life?

Sometimes, in fact often, the line between use and abuse is very fuzzy. Specific biological, ethical, social, and task-oriented impacts are often the only ways we have of judging whether or not the line has been crossed.

What to Do About Substance Abusers at Work

We had a lengthy discussion on this topic with Robert Driscoll, a psychotherapist in private practice in New York. He has had extensive experience counseling people and studying problems related to addictions. According to Driscoll, it is a human response to have concern for a co-worker you believe is in trouble, but what to do about it is not clear-cut. Most people are unwilling to confront someone with a drug or drinking problem until it's not possible to ignore it. At work, it's almost unthinkable—and very uncomfortable—to say something to a colleague about his or her drinking, snorting, smoking, free-basing, mainlining, or popping. Especially if it's the boss. We all make the distinction between a private and a professional life; the social taboo against invading someone else's private territory is very strong.

Because the use of street drugs is virtually illegal, taking these drugs at work is done covertly, although in many firms where to cop is known by everybody who's interested. Drinking alcohol, on the other hand, is legal and socially sanctioned; in fact, the workplace (i.e., the employer) might be considered to contribute to one's alcoholic drinking. "TGIF" is a call to the institution of belting down a few at the end of the week; martini lunches are a favorite institution (although not as much as formerly), as are the office party and the employees' picnic.

If a person you work with has a drinking or drug problem, or both, and you suspect or observe it, you need advice about what to do. If you do nothing, you contribute to a cover-up. Most well-meaning intentions don't work, as it happens, because dealing with addiction requires special knowledge that most people don't have.

Your firm may have an Employee Assistance Program; with or without an EAP, the firm may have a written policy for handling the growing number of employees who abuse drugs and alcohol. Find out what the policy is, and follow it. If there is no policy, and you want to help a co-worker (or to correct an increasingly intolerable work situation because somebody is clearly in trouble), you will do best to talk to your boss—or, if he or she is the person in trouble, to his or her boss—about what is wrong and what the firm is willing to do to assist the addicted employee. The employer has a range of options, from doing nothing, to warning the employee to shape up, to offering information about treatment and counseling, to requiring that the worker have treatment for the problem, to referring the employee to the Employee Assistance Program, to arranging an intervention (or confrontation) with the employee, his family, friends, and fellow workers under the direction of a trained drug addiction and alcoholism treatment specialist. This technique is used in the work setting to encourage a person to get help in improving his work performance and to keep from getting fired.

To talk to somebody about his or her addiction, you have to be very clear and precise. Substance abusers typically rely on denial to minimize what they are doing; primarily, they are hiding the truth from themselves; and what they do know about themselves, they don't believe anyone else notices. Often, they are the last ones to recognize the gravity of their morbid condition. Write down what you observe: date, time, rank and serial number. Be very clear and specific; don't diagnose.

Although a person may not come to work drunk or high, there are clear signs that he or she is drinking or using drugs heavily. A once routine person will become erratic in behavior; there may be unaccountable and frequent mood swings in somebody who is normally placid; a reliable person will become spasmodic in his work. The chronic substance abuser will be late for work and frequently absent, sometimes without explanation; he or she will tell lies, often obviously; the output of his or her work will decrease; accidents will increase; personal hygiene and grooming will deteriorate; the worker may start to avoid others; there may be drinking in the office, or getting high. Different drugs affect us differently. Although a drunk may have slurred speech, a telltale breath, stumbling gait, and impaired coordination, someone high on cocaine, for example, shows none of these signs; he or she is likely to be talkative and friendly; there may be no indication of trouble until there is an emergency.

The Prospects

In apparent contradiction to a resigned acceptance of substance abuse, there appears to be what *The New York Times* called "a new temperance taking root in America." During the eighties, alcohol consumption decreased gradually but steadily, particularly hard liquor, use of which declined approximately 25 percent. Although statistics about illicit drug use are, understandably, harder to come by, it seems that drug abuse may have declined—or, at the very least, leveled off—in recent years.

Within each of us, there seems to be a constant conflict between self-indulgence and self-control. Many observers feel that the post–World War II era was one in which self-indulgence got the upper hand in America. Since human behavior seems to swing like a pendulum from side to side, we may be moving into an era of reaction away from excessive self-indulgence and toward self-discipline. If this is so, the problems of substance abuse in the workplace may diminish as this new era takes hold.

All that assumes, however, that substance abuse is mostly, if not entirely, a behavioral problem. If, as seems more likely now, it is both a biological and a behavioral problem, the answer will not be so simple. Biologically based behavior may be influenced by community values at any given time, but the degree to which this is so is debatable. Furthermore, there is no guarantee that the new temperance, if it is real and substantial, will not at some point in the near or distant future fade like other trends.

The ultimate answer (or answers) will probably come from advances in brain mapping and genetics. As noted above, new and safe mood-altering substances may soon emerge from laboratories. Advances in brain and genetic mapping and testing may enable us to identify and isolate the signals or genes that affect susceptibility to or greater need for mood-altering substances. From research on the brain, we may find natural and less harmful ways of allowing our emotions to get a lift or a vacation. And genetic engineering could give us the means to change our genes and thereby redirect or eliminate the flaw of addiction.

And then, caffeine-free, tobacco-free, alcohol-free, and cocaine-free, we will face the office and our co-workers ready to conquer all tasks placed before us with clarity of mind and strength of character. Well . . . eliminating addictions can't solve *every* problem.

12 *Getting to Know You: How Private Is Your Biology?*

Because of the rapid pace of development of information technology, we are fast approaching the showdown at the OK Corral: Is it okay to increasingly invade personal privacy for the sake of workplace efficiency? Is it okay to use all the knowledge that technology is giving us to monitor and screen any or all workers? Is it okay to use personal employee data to keep employer health care costs down? Is it okay to monitor workers without their knowledge or permission? Is it okay to buy, sell, distribute, or collect personal health and life-style data on employees, co-workers, executives, or possible hires?

In *The Electronic Sweatshop*, by Barbara Gerson, we see how computers are increasingly being used for employee surveillance. She cites a U.S. Department of Labor estimate as far back as 1984 that nearly two-thirds of the people who worked at video display terminals were monitored by their employers. Tower Systems International is a company that promotes an E-mail monitoring program, which can not only track all attempted security violations, but also all electronic mail messages sent and received, and all sent but not yet opened.

It has many competitors; the field is growing rapidly.

Many employers believe that even monitoring your phone conversations is not illegal or unethical. And while legislation is under consideration in a number of states on this issue, individually purchased eavesdropping devices are being sold on the Main Streets of every city and town, ensuring that even if some official channels of surveillance are outlawed, others will be in place and undetectable. Since such monitoring is largely done in the service sector, few of you affected by it are covered by unions to represent your concerns.

(Even on a seemingly less intrusive basis, we find technology causing privacy problems that are largely felt but ignored. The Conference Board magazine, *Across the Board*, recently reported on a survey of two hundred executives, almost half of whom claimed a dislike for the speaker or conference phone system—on the basis of its intrusion into their privacy, as well as poor quality of communication that they felt resulted.)

The Shadow Knows

A University of Illinois study of *Fortune* 500 companies found that over half use private investigative agencies to get or check employee information, and about two-fifths of these get the information without telling the employee. Who has access to this information besides the company? Over a third of the companies have no policy with regard to releasing the information they've collected to the government. And 80 percent would give the data to an employee's creditor without a subpoena!

Even more to the point of this book, half the companies surveyed say they use the medical data they have collected to make employment-related decisions. And at some of the companies, medical claims are read by corporate processors before being sent to the insurance company. In these cases, people in the personnel office get to see all the claims filed for medical and psychological care.

In the age of substance abuse, AIDS, white collar crimes,* and rising health care costs, employers seem pressed to violate employee privacy, supposedly for the good of the organization. From hair analysis (a drug-testing device that has gained in popularity despite being of questionable reliability) to lie detector tests, there is a mounting assault on the employee's body to determine the future cost and benefit of that employee's presence.

These are matters most of us are already somewhat familiar with. Less well-known, and in the long run far more crucial, is the rapid approach of the most revolutionary biological breakthrough of the modern era—the mapping of the human genome, or the cracking of the human genetic code. There is a mind-boggling amount of effort and resources now being invested in gaining complete understanding of the *3 billion base pairs* of the 50,000 to 100,000 genes we are each made up of. Once the riddle of the genome is solved, we may see the ability to predict and screen health traits on a scale we can hardly now imagine. Will you get heart disease? Or will you have a tendency toward heart disease? Will you be violent? Or will you be a quietly depressed individual? At what age will your teeth lose bone? Will your hair turn white? Will your arthritis become incapacitating? And at what age will your physical health cause your office productivity to suffer, or your health benefits to cost more?

Who has the right to get access to, and use, genetic information on any individuals, and particularly on employees, will be the basis for extensive and animated debate. Ethicists are already highly concerned. The National Institutes of Health (NIH) are planning to use only 1 to 3 percent of their enor-

*Estimates of loss to business as a result of internal theft are in the billions of dollars—some say from $40 billion to $200 billion a year.

mous budget for the human genome project to explore the social, legal, and ethical issues that could be involved. The Human Genome Initiative, a group working on these issues, expresses concern that discrimination in insurance, and ultimately in the workplace, will extend beyond those severely and chronically disabled to include those of us who may be labeled with genetic traits we didn't even know we had.

There are some who believe that the technology, whether we morally accept it or not, will be so powerful as to change the very nature of the reproductive process. These people feel that in the not-too-distant future, our ability to completely analyze the genetic makeup of an embryo would lead to a marketplace of "high grade embryos," almost casting children in the role of commodities.

Eugenics—breeding humans the way we do animals and plants—may be decades off. But there are many things happening today that both threaten and protect the boundaries of your biological privacy at work. In March of 1992, Wisconsin governor Tommy Thompson signed America's first law restricting genetic testing in the workplace. In Wisconsin, such tests cannot be used as a condition of employment, but they can be used in workers' compensation investigations and in considering whether you may be susceptible to or exposed to a potentially toxic workplace substance. In any case, no punitive action may be taken against you.

The Bureau of National Affairs has found that juries have been increasing the number and size of awards to employees who claim invasion of workplace privacy. In issues ranging from video monitoring to sexual harassment, from breaches of AIDS confidentiality to—ultimately—genetic testing and screening, issues of citizenship rights in the workplace are being explored and judged.

You are a living, breathing data bank. Your thoughts, actions, and health and well-being are all encoded inside the body that defines you. Science is constantly finding the ways to help reveal that data, with or without your knowledge or intention, to interested others. If this doesn't concern you, relax, sit back, kick off your shoes, and dream about the wonders of how the future office will operate with biologically perfectly screened and tuned employees. If this does concern you, look up your congressperson's name and address and drop a letter in the mail. It's more private than sending it electronically.

13 *Make Me: Mastering Motivation, Mood, and Morale*

A man walking along one evening sees a young boy, crying, on his hands and knees looking forlornly at the gutter under a streetlight. "What's wrong, son?" asks the man.

"I lost my quarter," sobs the boy.

The man offers to help and joins the boy in intently examining the gutter. After about ten minutes without success, he asks the boy, "Are you sure you dropped it here?"

"No, I dropped it there," says the boy, pointing across the street.

"Then why are you looking here?" asks the puzzled man.

"Because," replies the boy logically, "the light is better here."

That boy, and that approach to solving problems, is symbolic of people management in the workplace today. Much of human resource practice puts psychological theory and statistics together to treat all employees as if they were only one part of René Descartes's mind-body dichotomy—the mind.

People become easier to see this way, because we can manage as if that mind were the same for everybody and the same for each individual under any and all circumstances.

Smile: You're on Candid Biology

As we come to know that people are whole and distinctive individuals, Cartesian duality misleads us in our efforts to understand and come to terms with motivation, mood, and morale. Certainly, we are deeply affected by psychological factors. We react to working conditions; we respond to leadership; we can be influenced by our views of how other people treat us. We can be roused to enthusiasm, or we can be depressed or angered, by what our bosses do.

But we are also affected by internal biological factors. Very often, in fact, what may seem to be psychological—i.e., depression—can instead be physiological. When such is the case, no amount of motivation, no amount of carrot-or-stick management practices, will make much of a difference.

More and more, we are coming to question the underpinnings of both personality and behavior. What is called mood, for example, we have been treating as only a mental state. When we think someone is in a "bad mood," our tendency is to smile, joke, compliment, tease, or in some other way lighten the environment and, thereafter, his or her mood. The assumption is that all states of mood are mental and can be altered by an act of will on the part of the affected individual, with assistance, of course, from cheerful others.

Sometimes that works; sometimes it doesn't. And when it doesn't, we assume the person chooses to stay in a bad mood. But it is possible—indeed, likely—that the mood is biologically induced.

Many organizations, as we have discussed elsewhere in this book, are encouraging their employees to adopt and maintain healthy life-styles. A survey of *Fortune* 500 companies, done recently by the A. Foster Higgins consulting firm, showed that more than 30 percent of those firms had incentive programs to encourage healthier life-styles among their employees. The major reasons are economic: healthier employees mean lower health care costs; and healthier employees mean better productivity. But there is little evidence that companies are also linking good health to motivation and morale. And they should be.

There Are People and There Are People

Two studies done in the mid-eighties by the University of California showed that levels of job satisfaction appeared to have little if any relation to working conditions, pay, or job status. People who came into the job with "happy dispositions" were less likely to be dissatisfied at work than those with "unhappy dispositions." As some wise person once said, you can't make a pessimist happy with good news.

This doesn't mean, of course, that employers should abandon concerns about job satisfaction and employee morale. What it does mean is that they need to develop greater understanding of the individual biological aspects of mood, motivation, and morale.

Psychologist Michael Maccoby, author of the best-selling *The Gamesman*, describes a category of managers and professionals he calls "self-developers." These are people who resist being controlled or manipulated and who, therefore, are largely unaffected by current forms of motivation. They are relatively indifferent to pay and status. They work to the beat of their own individual drummer.

What makes these people different? Maccoby, being a psychologist, sees them in terms of psychological makeup. But is that all? Are they also, as seems likely, biologically different?

We do know that creative people are indeed biologically different. Studies at McLean Hospital in Massachusetts demonstrate that enhanced creativity may be a positive characteristic associated with an inherited liability for manic depression. Companies these days are spending millions of dollars hiring creativity consultants and bringing in elaborate creativity-training programs. This is being done on the assumption that everyone can be taught or encouraged to be creative. But maybe all companies need to do is watch for people who are subject to mood swings (a characteristic associated with, but not exclusive to, manic depression) and rely on those people for creativity.

(It probably would be more fruitful, too, if companies would rethink their assumptions about why they don't have enough creativity. After working for more than twenty years with a great number and variety of organizations, we have concluded that all of them have available to them great untapped sources of creativity. In every case, organizational impediments prevent those sources from being seen and utilized. Pressures for conformity, risk aversion, fear of being wrong—all of the things that make up the bureaucratic mindset—stifle individual creativity. When a company says, in effect, that it wants everyone to look alike, it is also saying, even if unknowingly, that it wants everyone to think alike.)

A study conducted among faculty members at the University of Iowa's famous Writer's Workshop found that, over a fifteen-year period, 80 percent had experienced either severe depression or manic depression. Another study among psychiatric patients found that those with mild depression or pronounced mood swings were more likely to be creative artists—*or successful business executives!*

It has also been reported that creative people are more likely to have close relatives who are creative, leading to the inference that there is a genetic predisposition to creativity and, obviously, to depression.

Randy Larsen, a professor at the University of Michigan, believes that humans have not only appetite-regulation systems but also mood-regulation systems that are biological. And the mood-regulation systems, just like the appetite-regulation systems, strive to maintain equilibrium.

Some people, according to Professor Larsen, *need to add excitement* to their lives because their mood-regulation systems *reduce* reactions to external stimuli. For some others, the opposite is true: they *need a more peaceful environment* because their mood-regulation systems *amplify* or *augment* stimuli from the outside. This "augmenter/reducer" theory is based on the biology of personality and behavior.

Reducers, for example, find dull, routine work excruciating, and their performance in such jobs reflects that. Larsen gave a group of reducers two thousand math problems to solve; they did terribly—the dullness of the task adversely affected their intellectual skills.

Larsen estimates that reducers make up some 20 percent of the population. Traditional kinds of motivation will not greatly improve their performance on unexciting tasks. Probably the most cost-efficient thing to do would be to identify augmenters and give such jobs to them.

Brains, Bellies, and Bed Sizes

As we have written elsewhere in this book, there are many nonpsychological factors that can affect people's moods and behavior in the workforce—lighting, aromas, etc. These factors illuminate other paths that may give us a better understanding of why people do what they do, and why what management does so often fails to produce the desired results.

Recent research shows that moods can be altered through exercise or through eating certain foods, such as candy bars. These activities bring about chemical changes in the brain, and it is these changes that change mood.

There have been many studies—over a thousand so far—that demonstrate conclusively how people's moods improve as a result of exercise. Depression, anxiety, stress, and irritation can all be reduced or even eliminated. People can become calmer and more easygoing. According to Jeff Zwiefel, director of the National Exercise for Life Institute, exercise results in more efficient cardiovascular and respiratory systems. This, in turn, results in greater oxygen flow, which revitalizes the cells.

Most important, exercise helps the body make and release the chemicals that can create a sense of well-being, particularly endogenous opioids (mood enhancers) and beta-endorphins (natural pain killers).

Brain chemistry may be the key to more knowledge about human behavior and the important areas of morale and motivation. Norepinephrine, a chemical "messenger" found in the brain, produces a by-product, MHPG, that shows up in relatively high concentration in the urine of people who feel depressed and helpless. The unanswered question, of course, is which came first, the chicken or the egg? Does a high level of MHPG cause feelings of depression and helplessness, or do these feelings trigger the brain to produce more MHPG? Scientists believe that high levels of stress, such as in the work environment, can cause both the chemical reaction and the psychological reaction. Clearly, more tests are needed.

It is also becoming clearer that the differences in brain chemistry in people can be a major factor in how they respond to work. Reducing stress may raise morale for some people, but for others—reducers, for example—a lessening of stress could make the work environment intolerably dull.

In ancient Greek mythology, there was an innkeeper, Procrustes, who had only one bed in his inn. If any travelers did not fit the bed, Procrustes altered them: travelers who were too short were stretched, and too tall travelers were shortened (you don't want to know how). Philosophers use the term "Procrustean bed" to label any effort to alter reality in order to make it fit a preconception or theory.

Much of today's motivational management would delight Procrustes. Most research in this field is based on the idea that behavior is, to quote one authority, "a function of ability and motivation." What follows from this is the belief that you can get a desired result in the workplace by encouraging, forcing, or otherwise manipulating workers' behavior along desirable paths. The underlying belief is that all human activity in the workplace is conscious and goal-oriented.

While academics and executives admit that there can be differences among workers (in how they set objectives and how they respond to motivation), those differences are seen as mental and totally under the conscious control of the worker.*

All mind-centered approaches to controlling workplace behavior are controversial—perhaps increasingly so. Business and academic journals are full of disputes about various new, old, revived, and discredited theories of motivation. They are also full of case histories that show how Theory A enabled Company Z to reach new heights of productivity and morale or how Theory B destroyed Company Y. The authors of these case studies generally have a stake in the outcomes, so their objectivity can perhaps be questioned. Many of the other articles are so contrary to people's experiences in the real world that their credibility is in doubt.

All of these theories and articles assume a regularity and consistency in human behavior that just doesn't square with reality. Are you that constant and consistent? Do you do the same things the same way all the time—even under identical circumstances? If you do, you're a rare one.

The fact is, we're not just minds. Everything from our genetic makeup to the phases of the moon can and will affect

*It is ironic that while workplace behavior is seen as being based on individual will, there is a strong contrary view about behavior in society. The dependency or victimization approach says people's behavior is shaped by what was done to them in the past, and there are limits to their ability to overcome those past experiences.

our behavior. This truth is disturbing to many, because it implies that we know less and can do less about human behavior than we think. But it also indicates that as we do learn more about the inseparable mind and body of the whole person, we can come to a better understanding of why we do what we do—and what we can do about it.

You can, for example, alter your moods through diet and exercise. Techniques such as meditation and biofeedback can help you control your physiological as well as mental state. As you become more aware of and tuned into your own body, you can make yourself more efficient and effective.

The great mathematician and father of computer science, John von Neuman, said that some systems are so complex that any description of them would have to be more complex than the systems themselves. He was talking about human beings. It is foolish to believe that we can devise simple explanations for human behavior, when human behavior itself is so complex, variable, and ultimately unpredictable.

If you are a worker, and you don't respond to management's motivational efforts the way you are expected to or the way others do, don't just think that there is something wrong with you. Your individuality—in particular, your unique biology, physiology, and chemistry—combined with the circumstances of the moment, profoundly affects your response.

If you are a manager, recognize the limitations of all the theories and practices of human resource management. Individual moods and life-styles, genetic makeup, brain chemistry, weather, sensory input, and all the other factors we discuss in this book are such powerful influences that if you get the results you want, it may just be luck. Increasingly, in the future, good management will require better understanding of these factors. This is the true meaning of "managing diversity." It makes the job a lot more complicated—and difficult—but it may also result in more harmony and productivity in the office.

REFERENCES

INTRODUCTION

Blakeslee, Sandra. "Finding a New Messenger for the Brain's Signals to the Body." *The New York Times,* August 11, 1992, p. C3.

Edelman, Gerald M. *Bright Air, Brilliant Fire.* Basic Books, 1992.

Ezzell, C. "Monitoring Memories Moving in the Brain." *Science News,* May 2, 1991, p. 294.

———. "Morning Sickness Good for Baby." *Science,* August 7, 1992, p. 743.

Franklin, Deborah. "The Amazing Influence of Genes on Personality; Biological Traits Tempered by Environment." *Sacramento Bee,* September 10, 1989, pp. F1+ (reprinted from *The New York Times Magazine,* 1989).

Goode, Erica; Schraf, Joannie; and Burke, Sarah. "Where Emotions Come From." *U.S. News and World Report,* June 24, 1991, pp. 54+.

Hendricks, Melissa. "Neuroscientists Plan 'Atlas' of the Brain." *New Scientist,* February 9, 1991, p. 27.

Henig, Robin Marantz. "Body and Mind." *The New York Times Magazine,* April 17, 1988, pp. 59+.

Jaroff, Leon. *The New Genetics.* Whittle Direct Books, The Grand Rounds Press, 1991.

Kinoshita, June. "Mapping the Mind." *The New York Times Magazine*, October 18, 1992, p. 44+.

Mathison, G. Stewart. " 'Brain Dominance' Emerges as a New Marketing Research Technology." *Marketing News*, September 29, 1988, p. 13.

Naunton, Ena. "The Body of Emotions: Your State of Mind May Hold the Key to Physical Health." *Miami Herald*, September 13, 1989, pp. D1+.

Patlak, Margie. "Emotional Impact." *Los Angeles Times*, June 24, 1991, p. B3.

Plomin, Robert. "The Role of Inheritance in Behavior." *Science*, April 13, 1990, pp. 183+.

CHAPTER 1

Sight

Armstrong, Bruce. "Health Benefits of the Sun." *New Scientist*, November 19, 1988, p. 55.

Cushman, William H. and Christ, Brian. "Illumination," in *Handbook of Human Factors*, Gavriel Salvendy, ed. Wiley, 1987, pp. 671, 677–78, 680.

Flynn, J. E., et al. "Interim study of procedures for investigating the effect of light on impression and behavior." *Journal of the Illuminating Engineering Society* 3, pp. 87–94; quoted in Cushman and Crist, "Illumination," in *Handbook of Human Factors*, Gavriel Salvendy, ed., Wiley, 1987, pp. 679–80.

Harmon, D. B. "The Coordinated Classroom." Grand Rapids, MI, 1951, in Liberman, Jacob, *Light: Medicine of the Future*, Santa Fe, 1991, 55–57.

Harrigan, John E. "Architecture and Interior Design," in *Handbook of Human Factors*, Gavriel Salvendy, ed. Wiley, 1987, p. 743.

———. "Improving the quality of light . . . and work performance," in *Office Administration and Automation*. May 1984, pp. 38–

48; in Cushman and Crist, "Illumination," in *Handbook of Human Factors*, Gavriel Salvendi, ed., Wiley, 1987, p. 677.

Liberman, Jacob. *Light: Medicine of the Future*, Santa Fe, 1991, pp. 24, 25–26, 30–35, 58–59, 60–62, 121, 141–144.

Ott, J. N. "Color and Light: Their Effects on Plants, Animals and People." *Journal of Biosocial Research 7*, part I (1985), in Liberman, Jacob, *Light: Medicine of the Future*, Santa Fe, 1991, p. 57.

Robertson, A. S., et al. "Building Sickness, Are Symptoms Related to the Office Lighting?" *Annals of Occupational Hygiene*, vol. 33, no. 1, pp. 47–59.

Weale, R. A. Women's Occupational Health Resource Center, 1981.

———. "The Aging Eye," in *Handbook of Human Factors*, Gavriel Salvendy, ed. Wiley, 1987, p. 677.

Smell

Birchall, Annabelle. "A Whiff of Happiness." *New Scientist*, August 8, 1990, pp. 44 + .

Bishop, Jerry. "Biotech Researchers Follow a Siren Call: Human Pheromones." *The Wall Street Journal*, May 11, 1992, p. A-1.

Burton, Robert. *The Language of Smell*. Boston, 1976, pp. 108–15.

Bylinsky, Gene. "A Sixth Sense That Affects How You Feel." *Fortune*, January 27, 1992, p. 99.

The Fragrance Foundation, 145 East 32nd Street, New York, NY 10016.

Freedman, Alix. "Search Is on for Emotion-Eliciting Scents." *The Wall Street Journal*, October 13, 1988, p. B1.

———. "Feedback." *New Scientist*, July 18, 1992, p. 88.

Hainer, Cathy. "Aromatherapy Wafts into Mainstream Science, Business." *USA Today*, 1992.

Ingram, Jay. "The Science of Everyday Life." Penguin Books, 1987, pp. 140 + . From Gustavson, Dawson, and Bonett, "Androstenol, a

Putative Pheromone, Affects Human [Homo Sapiens] Male Choice Preference." *Journal of Comparative Psychology*, vol. 101, no. 2, 1987, pp. 210+.

Johnson, Dirk. "Therapists Offer Relief by Putting Patient on the Right Scent." *The New York Times*, November 8, 1988.

Lawlor, Julia. "Eau de Work." *USA Today*, 1992.

Lawren, Bill. "Beach Heads." *Omni*, September 1988, p. 28.

———. "On the Scent of a Better Day at Work." *New Scientist*, March 2, 1991, p. 18.

Luscher, Dr. Max. "The Luscher Color Test." Translated and edited from the German text by Ian A. Scott. New York: Random House, pp. 9–16.

O'Neill, Molly. "Taming the Frontier of the Senses: Using Aroma to Manipulate Moods." *The New York Times*, November 27, 1988, pp. C1+.

Solo, Sally. "What's Up—And Down—in Japan: Scents to Make You Work Harder." *Fortune*, February 27, 1992, p. 8.

Sound

Ingram, Jay. "Report to the President and Congress on Noise." Environmental Protection Agency, 92nd Congress Document No. 92–63, February 1982; quoted in Albert Thurmann and Richard E. Miller, *Fundamentals of Noise Control Engineering*, Atlanta, 1986, p. 1.

———. "Self-Reported Hearing Loss Among Workers Potentially Exposed to Industrial Noise–United States," JAMA *(Journal of the American Medical Association)*, April 15, 1988, pp. 2213+.

———. "The Science of Everyday Life." Penguin Books, 1989, pp. 59+. From MacLean, William R., "On the Acoustics of Cocktail Parties," *Journal of the Acoustical Society of America*, January 1959, pp. 79+.

Kingman, Sharon. "Quiet! Toxic Chemicals at Work." *New Scientist*, August 29, 1992, p. 15.

Suter, Alice. "Noise Wars." *Technology Review*, November-December 1989, pp. 42+.

———. "Quiet!! Save a Nation's Eardrums." *The Washington Post*, Sunday, November 12, 1990.

———. "VDT 'Noise' Causes Stress." *Issues in HR*, January 1991, p. 11.

Total Environment

Holden, Constance. "Controlling Your Personal Environment." *Science*, May 15, 1992, p. 963.

Sprout, Alison. "Products to Watch: Atmosphere Manager," *Fortune*, July 27, 1992, p. 107.

CHAPTER 2

Alstetter, Billy. "Long Days, Short Tempers." *Omni*, September 1992, p. 28.

Bower, Bruce. "Here Comes the Sun." *Science News*, July 25, 1992, pp. 62–63.

Bishop, Jerry E. "Study Finds Night-Shift Workers More Likely to Have Miscarriages." *The Wall Street Journal*, January 11, 1993, p. A5B.

Brody, Jane. "Doses of Pineal Gland Hormone Can Reset Body's Daily Clock." *The New York Times*, November 3, 1992, pp. C1+.

Drexler, Madeline. "Light Motif." *Boston Globe*, December 9, 1990, Magazine Section, pp. 42–43.

"Eye Sensitivity May Be at Fault in Seasonal Depression." *The New York Times*, Tuesday, May 12, 1992, p. C5.

"A Gallup Study on Workers' Conception of Productivity," July 1991. Prepared for Accountants On Call, Saddle Brook, New Jersey 07662.

Grossman, John. "Season of Our Discontent." *US Air Magazine*, November 1990, pp. 114+.

Lang, Susan. "Perform at Your Peak". *New Woman*, November 1992, pp. 100+.

McEnroe, Jennifer. "Split-Shift Parenting." *American Demographics*, February 1991, pp. 50+.

National Sleep Foundation. 122 South Robertson Blvd., Suite 201. Los Angeles, CA 90048 213-288-0466.

Rifkin, Jeremy. "20th Anniversary Gossip." *Whole Earth Review*, Winter 1988, p. 106.

Shull, Glen. "Prioritize Work According to Your Body Clock." *Boardroom Reports*, May 15, 1992, p. 13.

Smyth, Angela. "Seasonal Affective Disorder." Harper Books, 1992.

"They're Not Kidding When They Say 'Blue Monday.'" *Business Week*, May 18, 1992, p. 113.

Willis, Judith Levine. "Keeping Time to Circadian Rhythms." *FDA Consumer*, July/August 1990, pp. 19–21.

Young, Michael. "The Metronomic Society: Natural Rhythms and Human Timetables." Harvard University Press, 1988.

———. "Monday, Monday . . . Can't Work That Day." *Supervision*, June 1992, p. 26.

CHAPTER 3

Angier, Natalie. "Researchers Find a Second Anatomical Idiosyncrasy in Brains of Homosexual Men." *The New York Times*, August 1, 1992, p. 7.

Blakeslee, Sandra. "Female Sex Hormone Is Tied to Ability to Perform Tasks." *The New York Times*, November 18, 1988, pp. 1+.

———. "Why Don't Men Ask Directions? They Don't Feel Lost." *The New York Times*, May 26, 1992, pp. C1 and C8.

——— "DES Sons Show Changes in Brain Function." *Science News*, November 7, 1993, p. 318.

Facelmann, K. A. "Working Women Take One Benefit to Heart." *Science News*, June 24, 1989, p. 389.

———. "Working Women Accrue Heart Benefits." *Science News*, February 15, 1992, p. 141.

Fermiel, Stephen and Lublin, Joann. "Ruling on Sex Bias Curbs Employers." *The New York Times*, October 4, 1988, p. A25.

———. "Justices Bar 'Fetal Protection' Policies." *The Wall Street Journal*, March 31, 1991, pp. B1+.

Friedman, Dana. "Why the Glass Ceiling?" *Across the Board*, July/August 1988, pp. 33+.

Gibbons, Ann. "The Brain as Sexual Organ." *Science*, August 30, 1991, pp. 957–59.

Hermann, Ned. *The Creative Brain*. North Carolina: Brain Books.

Holden, Constance. "Gender Spatial Differences Are Zero." *Behavior Today*, March 13, 1989, pp. 7+.

———. "Is 'Gender Gap' Narrowing?" *Science*, August 30, 1991, pp. 959–60.

Hughes, Kathleen. "Did You Know That . . ." *Boardroom Reports*, January 11, 1990, p. 15.

———. "Pregnant Professionals Face Pressures." *The Wall Street Journal*, February 6, 1991, pp. B1+.

Lewin, Tamar. "Companies Ignore Men's Health Risk." *The New York Times*, December 15, 1988, p. A24.

Moir, Anne and Jessel, David. *Brain Sex*. New York: Carol Publishing Group, 1991.

Morrison, Ann; Van Velsor, Ellen; White, Randall; and the Center for Creative Leadership. *Breaking the Glass Ceiling*. Reading, Mass.: Addison Wesley Publishing Group, 1987.

"New Woman and NAFE Announce Results of the 1992 New-Business Contest Survey: Women Tell Why They Want to be

Their Own Boss." *New Woman Magazine* and The National Association for Female Executives, October 8, 1992, press release.

O'Brien, Jim. "It's That Time of Year." *Omni*, August 1992, p. 28.

Otten, Alan. "Male Hormone Linked to Marital Problems." *The Wall Street Journal*, August 19, 1992. p. B1.

Poe, Randall and Baker, Emily. "Working Mothers." *Across the Board*, July/August 1990, p. 3.

Roberts, Leslie. "British Radiation Study Throws Experts into Tizzy." *Science*, April 6, 1990, pp. 24+.

Schmidt, William. "Risk to Fetus Ruled as Barring Women from Jobs." *The New York Times*, October 3, 1989, p. A16.

Skerrett, P. J. "Blame It on Your Father." *Omni*, September 1991, p. 36.

Tavris, Carol. *The Mismeasure of Woman*. New York: Simon & Schuster, 1992.

Tear, Jayne, president of The Jayne Tear Group, 225 West 71st Street, New York, NY 10023, 212-580-4100.

———. "Gender Gap! Do Women Sell Themselves Short in Performance Reviews?" "Labor Letter," *The Wall Street Journal*, July 21, 1992.

Wickelgren, I. "Drinking Problems Rise Among Young Women." *The New York Times*, October 13, 1988, p. D10.

———. "Firming the Figures on Mental Illness." *Science News*, November 12, 1988, p. 311.

Woo, Junda. "Gender Segregation Prominent in Elementary School." *Behavior Today*, January 11, 1988, p. 7.

———. "Feminist Legal Theory Enters Business Arena." *The Wall Street Journal*, June 4, 1991, pp. B1+.

CHAPTER 4

Angier, Natalie. "A Male Menopause? Jury Is Still Out." *The New York Times,* May 20, 1992, p. C14.

Arthur, Charles. "Design for the Third Age." *New Scientist,* June 13, 1992, pp. 24T.

Banner, Lois W. "Future of Retirement Symposium Proceedings." Society of Actuaries, 1988.

———. *In Full Flower* New York: Alfred A. Knopf, 1992.

———. "What Lies Ahead." United Way of America, 1992.

Brody, Jane E. "Can Drugs 'Treat' Menopause?" *The New York Times,* May 20, 1992, p. C14.

———. "Alternatives to Hormone Therapy After Menopause." *The New York Times,* May 20, 1992, p. C14.

Coates, Joseph F.; Jarratt, Jennifer; and Mahaffie, John B. "Future Work." *The Futurist,* May/June 1991, pp. 9+.

Deyo, R. A.; Straube, K. T.; and Disterhoft, J. F. "Nimodipine Facilitates Associate Learning in Aging Rabbits." *Science,* February 10, 1989, pp. 809+.

Fisher, Helen E. "Mighty Menopause." *The New York Times,* October 21, 1992, p. A23.

Foley, Patrick and Moray, Neville. "Sensation, Perception, and Systems Design," in *Handbook of Human Factors,* Gavriel Salvendy, ed. Wiley, 1987, pp. 45+.

Goldspink, Geoffrey. "A Is for Eyes." *Prevention,* February 1989, p. 8.

———. "The Brains Behind the Brawn." *New Scientist,* August 1, 1992, pp. 28+.

Gutfeld, Greg. "Memory Pills—Finally!" *Prevention,* August 1991, p. 41.

Jennings-Saver, Cheryl. "The Carrot Connection." *New Choices for the Best Years,* November 1990, p. 40.

O'Donnell, Michael. "Fading Memories." *International Management,* October 1991, p. 103.

Ostroff, Jeff. "An Aging Market." *American Demographics*, May 1989, pp. 26+.

———. "Menopause." *Newsweek*, May 25, 1992, pp. 71+.

Prentice, Thompson. "Slowing Down the March of Time." *World Press Review*, February 1989, pp. 31+.

Sheehy, Gail. *The Silent Passage*. New York: Random House, 1992.

Small, A. M. "Design for Older People," in *Handbook of Human Factors*, Gavriel Salvendy, ed. Wiley, 1987, pp. 495+.

Stone, Cathy. "Artificial Senility." *Omni*, July 1988, p. 29.

CHAPTER 5

Beasley, Joseph D., M.D. *Betrayal of Health: The Impact of Nutrition, Environment, and Lifestyles on Illness in America*, Times Books, 1991.

Bishop, Katherine. "Researchers Probing Feasibility of 'Designer Foods.' " *Technology Forecasts*, April 1992, p. 10.

———. " 'Smart Drugs': Elixir or Snake Oil?" *The New York Times*, June 10, 1992, pp. C1+.

———. "Interview: Sarah Leibowitz." *Omni*, May 1992, pp. 73+.

———. "Can a Nutrient in Milk Help Save the Brain from Stroke Damage?" *Business Week*, July 16, 1992, p. 65.

Braus, Patricia. "The Bottom Line." *American Demographics*, October 1989, pp. 34+.

———. "Wellness Incentive." *Board Room Reports*, June 15, 1991.

Brody, Jane E. "Curbing Senior Risks." *The New York Times*, May 20, 1992, p. C14.

Burros, Marian. "Vitamins Are Fine, With the Right Foods." *The New York Times*, June 24, 1992, p. C3.

Chinen, Deane. "The Moral Implication of Nutritional Therapy." *The Humanist*, January-February 1988, pp. 26+.

Hall, Trish. "Pass the Spelt . . ." *The New York Times*, June 24, 1992, pp. C1+.

Kessler, David A., et al. "The Safety of Foods Developed by Biotechnology." *Science*, June 26, 1992, pp. 1747+.

Noble, George. "Vitamins and I.Q." *Omni*, July 1988, p. 32.

Pearce, Fred. "Soils Spoilt by Farming and Industry." *New Scientist*, May 16, 1992, p. 7.

Pereira, Joseph. "Vitamin-Mineral Supplements Improve Student Ability." *FutureScan*, February 18, 1988, pp. 3+.

———. "The Exercise Boom Loses Its Strength." *The Wall Street Journal*, January 9, 1989, pp. 1+.

———. "Another 'F.'" *American Demographics*, November 1989, pp. 18+.

Solomon, Dr. David H., et al. *A Consumer's Guide to Aging*. The Johns Hopkins University Press, 1992.

Stipp, David. "Science Is Pushing Heart Disease's Toll Further into Old Age." *The Wall Street Journal*, January 24, 1990, pp. A1+.

———. "Lots of Oat Bran Found to Cut Cholesterol." *The New York Times*, June 24, 1992, p. C14.

CHAPTER 6

Aldhous, Peter. "Companies Teaching Workers 3Rs to Compete in Age of High Technology." *The New York Times*, May 10, 1988, p. 26.

———. "The Promise and Pitfalls of Molecular Genetics." *Science*, July 10, 1992, pp. 164+.

———. "Scan Reveals Musician's Skill." *New Scientist*, July 11, 1992, p. 7.

Barinaga, Marcia. "Knockouts Shed Light on Learning." *Science*, July 10, 1992, pp. 162+.

———. "The Brain Remaps Its Own Contours." *Science*, October 9, 1992, pp. 216+.

Barzun, Jacques. "Television and the Schools—But Not What You Think." *Basic Education Issues, Answers and Facts*, Fall 1987.

Blakeslee, Sandra. "Behavior Therapy Can Change How the Brain Functions, Researchers Say." *The New York Times*, September 16, 1992, p. A20.

Desimone, Robert. "The Physiology of Memory: Recording of Things Past." *Science*, October 9, 1992, pp. 245+.

Edelman, Gerald M. *Bright Air, Brilliant Fire*. Basic Books, 1992.

Erlich, Julie. "Brain Gain: Drugs That Boost Intelligence." *Omni*, September 1992, pp. 42+.

Fiske, Edward B. "In Indiana, Public School Makes 'Frills' Standard." *The New York Times*, May 24, 1988, p. A16.

Flaste, Richard, ed. *The New York Times Book of Science Literacy*. Harper Collins, 1992.

Gold, Karen. "If All Else Fails, Read the Instructions." *New Scientist*, June 13, 1992, pp. 38+.

Holden, Constance. "Academy Helps Army Be All That It Can Be." *Science*, December 11, 1987, pp. 1501+.

Hooper, Judith and Teresi, Dick. *The 3-Pound Universe*. Jeremy P. Tarcher Inc., 1986.

Kyllonen, P. C. and Alluisi, E. A., "Learning and Forgetting Facts and Skills," in *Handbook of Human Factors*, Gavriel Salvendy, ed. Wiley, 1987, pp. 124+.

Livermore, Beth. "Build a Better Brain." *Psychology Today*, September/October 1992, pp. 40+.

Main, Jeremy. "Trying to Bend Managers' Minds." *Fortune*, November 23, 1987, pp. 95+.

———. "Team Work Helps Fill in Missing Skills." *Issues in HR*, January 1991, p. 4.

Milunsky, Aubrey, M.D. *Heredity and Your Family's Health*. Johns Hopkins University Press, 1992.

Penrose, Roger. *The Emperor's New Mind.* Oxford University Press, 1989.

Poe, Randall, et al. "American Companies Accommodate." *Across the Board,* December 1990, p. 6.

Rose, Frederick. "An 'Electronic Clone' of a Skilled Engineer Is Hard to Create." *The Wall Street Journal,* August 18, 1988, pp. 1+.

Shanker, Albert. "Exploring the Missing Connection." *The New York Times,* June 19, 1988, p. E7.

———. " 'Functional Illiterates' Can Read! And Do a Job in Business and Industry." *The New York Times,* October 30, 1988, p. 7.

CHAPTER 7

Alexander, David. *The Practice of Industrial Ergonomics.* Englewood Cliffs, NJ, 1986, pp. 9, 66–67.

Brody, Jane. "Epidemic at the Computer: Hand and Arm Injuries." *The New York Times,* March 3, 1992, pp. C1+.

———. "Computer Users' Injuries Are Often Preventable." *The New York Times,* March 4, 1992, p. C14.

Burson, Rick. "The Economics of Ergonomics." *Ohio Monitor,* February 1991, pp. 9+.

Chaffin, Don. "Biomechanical Aspects of Workplace Design," in *Handbook of Human Factors,* Gavriel Salvendy, ed. Wiley, 1987, pp. 608, 615.

Davis, Louis and Wacker, Gerald. "Job Design," in *Handbook of Human Factors,* Gavriel Salvendy, ed., Wiley, 1987, p. 433.

Elliott-Proctor, Brenda. "Ergonomics, Anyone." *Ohio Monitor,* December 1991, pp. 5+.

Fognascher, Jim. "A Sound Approach to Ergonomics: Ergonomics in Workplace Modification, Not Worker Adaptation." *Ohio Monitor,* December 1991, pp. 10+.

Grandjean, Etienne. "Design of VDT Workstations," in *Handbook of Human Factors,* Gavriel Salvendy, ed. Wiley, 1987, p. 1360.

Habes, Daniel and Putz-Anderson, Vern. "The NIOSH Program for Evaluating Biomechanical Hazards in the Workplace." *Journal of Safety Research,* Summer 1985, pp. 49+.

Hecht, Jeff. "Telephone Network Brings Hope for the Deaf." *The New Scientist,* December 10, 1987, pp. 30+.

Horowitz, Janice. "Crippled by Computers." *Time,* October 12, 1992, pp. 70+.

Kanter, Rosabeth Moss. *Men and Women of the Corporation.* New York, 1977, p. 207.

Keyserling, William; Armstrong, Thomas; and Punnett, Laura. "Ergonomic Job Analysis: A Structured Approach for Identifying Risk Factors Associated with Overexertion Injuries and Disorders." *Appl. Occup. Environ. Hyg.,* May 1991, pp. 353+.

Kilhorn, Peter. "Workplace Injury Is Rising, and the Computer Is to Blame." *The New York Times,* November 16, 1989, p. A26.

Kochenburger, Linda. "Women in Workforce Still Hindered by Ergonomic and Equipment Problems." *News Digest,* August 1987, pp. 1+.

Kroemer, Karl. "Engineering Anthropometry," in *Handbook of Human Factors,* Gavriel Salvendy, ed. Wiley, 1987, pp. 161, 166.

Lewis, Peter. "An Awareness of Disabled Workers." *The New York Times,* April 30, 1989, p. 12.

Meister, David. "System Design, Development and Testing," in *Handbook of Human Factors,* Gavriel Salvendy, ed. Wiley, 1987, p. 18.

NASA. *Anthropometric Source Book.* 3 vols. Houston, Texas, 1978.

Parmeggiani, Dr. Luigi, tech. ed. *Encyclopedia of Occupational Health and Safety.* Vol. 1. Geneva: International Labor Office, pp. 775–778.

Rigdon, Joan. "How a Plant Handles Occupational Hazard with Common Sense—Many Small Changes Enable Sara Lee Bakery to

Ease Carpal Tunnel Syndrome." *The Wall Street Journal*, September 28, 1992, pp. A1+.

Rohmert, Walter. "Physiological and Psychological Work Load Measurement and Analysis," in *Handbook of Human Factors*, Gavriel Salvendy, ed. Wiley, 1987, p. 412.

Schachter, Jim. "Ergonomics." *Los Angeles Times*, June 27, 1988, part IV, p. 5.

Schneidawind, John. "High Voices Take #-ing." *USA Today*, 1992.

Smith, Jeff. "It's All in the Wrist." *Maine Telegram*, April 1, 1990, pp. 1C+.

Smith, Leo and Smith, James. "Human Factors Engineering Checklist," in *The Practice of Industrial Ergonomics*. Englewood Cliffs, N.J., 1986, p. 57–59.

Steelcase Inc. *Steelcase Worldwide Office Environment Index, 1991 Summary Series.*

Warldrop, Judith. "From Handicap to Advantage." *American Demographics*, April 1990, pp. 33+.

Wasil, James. "Ergonomics and Human Factors Seek Solutions to Old Workplace Problems." *Monitor*, May 1984, pp. 5+.

CHAPTER 8

Allen, Frank Edward. "Printers Are Injecting Ozone into Offices." *The Wall Street Journal*, January 29, 1992, p. B1.

Altman, Lawrence K. "Pregnant Women's Use of VDTs Is Scrutinized." *The New York Times*, June 5, 1988, p. 22.

Anderson, Ian. "Getting to the Heart of the Chinese." *Science News*, March 11, 1989, p. 156.

———. "Judge Backs Case Against Passive Smoking." *New Scientist*, February 16, 1991, p. 14.

———. "What Is Making These People Sick?" *Chemecology*, February 1991, pp. 2–3.

Becher, Robert O. "Views on Biology." *USACOR Newsletter*, March 1988, pp. 7+.

Blakeslee, Sandra. "Cell Test Suggests Link Between Cables and Cancer." *New Scientist*, December 3, 1987, p. 28.

———. "Magnetic Crystals, Guides for Animals, Found in Humans." *The New York Times*, May 12, 1992, pp. C1+.

Coghlan, Andy and Hall, Nina. "How Magnetic Fields Could Upset Your Ions." *New Scientist*, August 4, 1990, p. 30.

Collingwood, Harris. "Why Men Should Also Think of the Baby." *New Scientist*, March 2, 1991, p. 16.

———. "Breast Cancer and PCBs: A Possible Link." *Business Week*, April 6, 1992, p. 36.

Coon, Stanley. "The Left-Hander Syndrome." The Free Press/Macmillan, 1992.

Crispell, Diane. "Workers in 2000." *American Demographics*, March 1990, pp. 36–40.

Dadd, Debra Lynn. *The Nontoxic Home and Office*. Los Angeles, CA: Jeremy B. Tarcher Inc., 1992.

Davis, Bob. "These Fellows May Be Bad News for All You Hay Fever Sufferers." *The Wall Street Journal*, June 3, 1992, p. B1.

Foss, Laurence and Rothenberg, Kenneth. *The Second Medical Revolution: From Biomedical to Infomedical*. Random House, 1987.

———. " 'Sick' Buildings Exert Stressful Impact." *Science News*, April 25, 1992, p. 260.

———. "On Healthy Indoor Air, and Avoiding Suits." *The New York Times*, May 1992.

Holzman, David. "Elusive Culprits in Workplace Ills." *U.S. News & World Report*, June 26, 1989, pp. 44–45.

Janson, Elisabeth. "Brain Damage." *World Press Review*, April 1990, p. 72.

Lyall, Sarah. "Sick Buildings Get Breath of Fresh Air." *Science News*, February 15, 1992, p. 109.

———. "An Indoor Plague in Albany." *The New York Times*, February 17, 1992, pp. B1 +.

Nair, Indira and Morgan, M. Granger. "Electromagnetic Fields: Biological Effects." *IEEE Spectrum*, August 1990, pp. 23 +.

Paul, Bill. "Men Exposed to Electromagnetic Fields in Study Have Slower Motor Responses." *The Wall Street Journal*, December 6, 1989, p. B4.

Raloff, J. "Greenery Filters Out Indoor Air Pollution." *Science News*, October 30, 1989, p. 212.

Saltzman, Amy and Silberman, Joanne. "When Each Day Is a Sick Day." *U.S. News & World Report*, March 13, 1989, pp. 65—67.

Stevens, William K. "Scientists Debate Health Hazards of Electromagnetic Fields." *The New York Times*, July 11, 1989, pp. C1 +.

Stone, Richard. "Immunology: Pollutants a Growing Threat." *Science*, April 3, 1992, pp. 28 +.

CHAPTER 9

Brand, Stewart. "Virtual World Sickness." *Whole Earth Review*, Summer 1989, pp. 86 +.

Broad, William J. "Science Can't Keep Up With Flood of New Journals." *The New York Times*, February 16, 1988, pp. C1 +.

Brody, Jane E. "Epidemic at the Computer: Hand and Arm Injuries." *The New York Times*, March 3, 1992; pp. C1 +.

Heims, Steven J. *The Cybernetics Group*. MIT Press, 1992.

———. "Computer Games." *Newsweek*, August 31, 1992, p. 10.

Hughes, Sylvia. "Pilots Demand European Safety Watchdog After Airbus Crash." *New Scientist*, February 1, 1992, p. 14.

Johnson, George. "Life as a Program on a Cosmic Machine." *The New York Times*, February 19, 1989, p. 20.

Kulade, Dennis. "Into the Void." *The New York Times*, January 12, 1988, pp. 1+.

Lucky, Robert W. "Reflections: What's Going On?" *IEEE Spectrum*, November 1990, p. 6.

Mason, Jack. "Integrating Both Sides of the Brain." *The West Side Spirit*, April 9, 1989, pp. 6+.

Trachtenberg, Jeffrey A. "Viewer Fatigue?" *Forbes*, December 26, 1988, pp. 120+.

Zuboff, Shoshana. *In the Age of the Smart Machine*, Basic Books, New York, 1984.

CHAPTER 10

Barnathan, Joyce. "The Pregnant Silence in Corporate America." *Business Week*, March 9, 1992, p. 12. (A review of *Breaking With Tradition*, by Felice Schwartz, Warner Books, 1992.)

Blumenthal, Robin Goldwyn. "Survey Finds 25 Percent of Work Force Has Anxiety Disorder, Stress-Linked Illness." *The Wall Street Journal*, October 20, 1989, p. B2E.

Bower, Bruce. "Emotional Stress Linked to Common Cold." *Science News*, August 31, 1991, p. 132.

Chandrasekhar, R. "Meditation Power." *World Press Review*, September 1992, p. 49.

Diegmueller, Karen. "Huddled Masses Yearning for Space." *Insight*, May 30, 1988, pp. 8+.

Fackelmann, Kathy. "Hostility Boosts Risk of Heart Trouble." *Science News*, January 28, 1989, p. 60.

Farnham, Alan. "Read This or We'll Cut Off Your Ear." *Fortune*, December 31, 1990, pp. 88+.

Fox, Barry. "Electric Gun Shock Worries Police." *The New Scientist*, February 25, 1988, p. 38.

———. "Secret Weapon." *The New Scientist*, March 31, 1988, p. 35.

Friedman, Dana. "Why the Glass Ceiling?" *Across the Board*, July/August 1988, pp. 33+.

Fuller, Jim. "Surviving Job Burnout." *Star Tribune*, May 15, 1988, pp. 1E+.

Griffiths, Joan. "The Mirthful Brain." *Omni*, August, 1992, p. 18.

Grover, Ronald. "Say, Does Worker Comp Cover Wretched Excess?" *Business Week*, July 22, 1991, p. 25.

Itoh, Yoshiaki. "Worked to Death in Japan." *Whole Earth Review*, March 1991, p. 50.

Karasek, Robert and Theorell, Tores. *Healthy Work: Stress, Productivity, and the Reconstruction of Working Life*. New York: Basic Books, 1990.

Karr, Albert. "Labor Letter," *The Wall Street Journal*, May 7, 1991, p. 1.

Kelly, Marjorie. "Revolution in the Marketplace." *The UTNE Reader*, January–February 1988, pp. 54+.

———. "Mediating for a Discount." *Hastings Center Report*, April/May 1988, p. 48.

Lefkoe, Morty. "Unhealthy Business." *Across the Board*, June 1992, pp. 26+.

McCarthy, Michael. "Stressed Employees Look for Relief in Worker's Compensation Claims." *The Wall Street Journal*, April 7, 1988, p. 31.

McCarthy, Paul. "Jam Sessions." *U.S. News and World Report*, September 7, 1987, pp. 20–27.

———. "Stressful Screens." *Omni*, August 1992, p.29.

Middlemist, Knowles, and Matter. "Personal Space Invasions in the Laboratory: Suggestive Evidence for Arousal." *Journal of Personality and Social Psychology*, vol. 33, no. 5, 1976, pp. 541–546. As discussed by Jay Ingram in *The Science of Everyday Life*, Penguin Books, pp. 133+.

———. "Law, Crime and Justice." *Future Survey*, March 1989, pp. 8+.

Naunton, Ena. "When You Can't Fight or Flee, You Can Still Cope With Stress." *Miami Herald*, September 13, 1989, pp. D1+.

———. The National Headache Foundation, Chicago, 1-800-843-2256.

Noble, Barbara Presley. "Round Two on the Mommy Track." *The New York Times*, February 23, 1992, p. F23.

Orski, C. Kenneth. " 'Managing' Suburban Traffic Congestion: A Strategy for Suburban Mobility. *Transportation Quarterly*, October 1987.

Robinson, John. "Time Squeeze." *American Demographics*, February 1990, pp. 30+.

Schwartz, Joe. "Future Lock." *American Demographics*, November 1989, p. 16.

Shellenbarger, Sue. "Measuring the Impact When Job, Family Collide." *The Wall Street Journal*, July 22, 1991, p. B1.

Smith, E.; Brott, Jody; Cuneo, Alice; and Davis, Jo Ellen. "Stress, the Test Americans Are Failing." *Business Week*, April 18, 1988, pp. 74+.

———. "Dysfunctional Families Produce Dysfunctional Managers." *Issues in HR*, January 1991, p. 8.

Stewart, Thomas. "Do You Push Your People Too Hard?" *Fortune*, October 11, 1990, pp. 121+.

Stone, Eden and Taylor, Peggy. "Japanese Fear They'll Be Worked to Death." *Issues in HR*, January 1991, p.13.

———. "The Overworked American: An Interview with Harvard Economist Juliet Schor." *New Age Journal*, November/December 1991, pp. 34+.

Tierney, John. "Burned Out: The People Who Really Suffer Job Stress." *Sacramento Bee*, May 15, 1988, pp. F2+.

Trost, Cathy. "Men, Too, Wrestle With Career-Family Stress." *The Wall Street Journal*, November 1, 1988, p. B1.

Ventura, Michael. "Dreamtime." *Whole Earth Review*, Winter 1988, pp. 2+.

Vobejda, Barbara. "When Your Job Makes You Sick." *Washington Post*, March 10, 1992, pp. 12+.

———. U.S. Air Shuttle, Advertisement, *The New York Times*, July 15, 1992, p. A19.

CHAPTER 11

Bower, Bruce. "Drugs of Choice." *Science News*, December 16, 1989, pp. 392+.

"Drug and Alcohol Abuse by Employees." *Across the Board*, July–August 1989, p. 6.

Fackelman, K. A. "Compulsive Gambling." *The Wall Street Journal*, June 21, 1988, p. 1.

———. "Male Teenagers at Risk of Steroid Abuse." *Science News*, December 17, 1988, p. 391.

Groves, David. "Steep Thrills." *American Health*, December 1988, pp. 71+.

Hall, Trish. "A New Temperance Is Taking Root in America." *The New York Times*, March 15, 1989, pp. A1+.

Holden, Constance. "Mind-Bending Drugs Could Leave Brains Permanently Warped." *New Scientist*, January 21, 1989, p. 30.

———. "Cocaine Shrinks Your Brain?" *Science*, March 13, 1990, p. 167.

Malcolm, Andrew H. "Crack, Bane of Inner City, Is Now Gripping Suburbs." *The New York Times*, October 1, 1989, pp. 1+.

Rosen, Robert H. *The Healthy Company*. New York: Tarcher/Parigee (Putnam), 1991.

Schuller, Zachary. "Adults Not Outdone by Their Children in Drug Abuse." *Behavior Today*, December 19, 1988, pp. 7+.

———. "If You Light Up on Sunday, Don't Come In on Monday." *Business Week*, August 26, 1991, pp. 68+.

Slomka, Jacquelyn. "Playing with Propranolol." *Hastings Center Report*, July–August 1992, pp. 13+.

Smothers, Ronald. "Employers Becoming Targets of Suits in the Fight to Halt Drunken Driving." *The New York Times*, December 24, 1991, p. A10.

———. "Smoking Declines at a Faster Pace." *The New York Times*, May 22, 1992, p. A17.

Wright, Robert. "Alcohol and Free Will." *The New Republic*, December 14, 1987, pp. 14+.

———. "Don't Protect the Alcoholic from Reality." *Behavior Today*, February 8, 1988, pp. 2+.

CHAPTER 12

Abramson, Jill. "Mind What You Say; They're Listening." *The Wall Street Journal*, October 25, 1989, p. B1.

Annas, George J. "Who's Afraid of the Human Genome?" *Hastings Center Report*, July 8, 1989, pp. 19+.

Blakeslee, Sandra. "Ethicists See Omens of an Era of Genetic Bias." *The New York Times*, December 27, 1990, p. D9.

Goode, Stephen. "Privacy Rights Now Aired in Public." *Insight*, February 2, 1988, pp 52+.

———. "Privacy." *Business Week*, March 28, 1988, pp. 61+.

Jaroff, Leon. *The New Genetics*. Whittle Direct Books, The Grand Rounds Press, 1991.

Karràker, Roger. "The Electronic Sweatshop." *Whole Earth Review*, Winter 1991, p. 68 (a review of the Barbara Gerson book, Penguin, USA, 1989).

Murphy, John A. "Whose Business Is It?" *Vital Speeches of the Day*, November 15, 1988, pp. 146+.

Nelkin, Dorothy and Tancredi, Laurence. *Dangerous Diagnostics: The Social Power of Biological Information*." New York: Basic Books, 1989.

Nolan, Kathleen and Swenson, Sara. "New Tools, New Dilemmas: Genetic Frontiers." *Hastings Center Report*, October/November 1988, pp. 40+.

Poe, Randall; Baker, Emily; and Courter, Carol Lee. "Fast Forward: Get Me Off That Squawk Box." *Across the Board*, December 1990, p. 6.

Slack, Nancy. "Biological Diagnostics: Boon and Bane." *Hastings Center Report*, January/February 1991, pp. 38+. Review of *Dangerous Diagnostics: The Social Power of Biological Information*, by Dorothy Nelkin and Laurence Tancredi.

Sloane, Leonard. "Just Who Can Raise the Shades and Peek into Private Matters?" *The New York Times*, January 25, 1992, p. 10.

Solomon, Jolie. "As Firms' Personnel Files Grow, Worker Privacy Falls." *The Wall Street Journal*, April 19, 1989, p. B1.

———. "Hair Analysis Unreliable?" *Issues in HR*, January 1991, p. 10.

Weinberg, Robert. "The Dark Side of the Genome." *Technology Review*, April 1991, p. 44.

Wertz, Dorothy and Fletcher, John C. "Disclosing Genetic Information: Who Should Know?" *Technology Review*, July 1989, pp. 22+.

———. "Wisconsin Bars Genetic Testing." *Issues in HR*, May–June 1992, p. 1.

CHAPTER 13

Bauer, B. "Mood Swings and Creativity: New Clues." *Science News*, October 24, 1987, p. 262.

Brown, B. "Manic Depression: Risk and Creativity." *New Scientist*, September 3, 1988, p. 151.

Buffington, Perry. "In the Mood." *Sky*, October 1989, pp. 102+.

Case, John. "Why Work?" *Inc*, June 1988, pp. 25+.

———. "Lifestyles of the Emotions." *Psychology Today*, January–February 1992, p. 10.

Hoyos, Carl G. "Motivation," in *Handbook of Human Factors*, Gavriel Salvendy, ed. Wiley, 1987, pp. 108+.

———. "Powerless Chemistry of Depression." *Science News*, June 13, 1992, p. 396.

Maccoby, Michael. "Self-Developers: Why the New Engineers Work." *IEEE Spectrum*, February, 1988, pp. 50+.

Reibstein, Larry. "Are Some Workers Bound to Be Unhappy?" *The Wall Street Journal*, April 18, 1988, p. 27.

INDEX

ABOUT THE AUTHORS

Edith Weiner and Arnold Brown are partners in the consulting firm Weiner, Edrich, Brown, Inc., and leading futurists. They have co-authored a column on the future for *Association Management*, and their work has been reported in hundreds of periodicals, including *The Wall Street Journal*, *Fortune*, *USA Today*, and *Planning Review*. Their book *Supermanaging: How to Harness Change for Personal and Organizational Success* was selected by three major book clubs and was published in Europe, Japan, and Latin America. They were the subject of a PBS television program on business planning, and have been consulted by *Business Week*, *The Wall Street Journal*, *The New York Times*, and *Fortune* for ideas about significant trends in business. Brown and Weiner are frequently asked to speak at conferences, on campuses, at corporate meetings, and on TV and radio.

Additional copies of *Office Biology* may be ordered by sending a check for $21.95 (please add the following for postage and handling: $2.00 for the first copy, $1.00 for each additional copy) to:

MasterMedia Limited
17 East 89th Street
New York, NY 10128
(212) 260-5600
(800) 334-8232
(212) 546-7607 (fax)

Edith Weiner and Arnold Brown are available for speeches and seminars. Please contact MasterMedia's Speakers' Bureau for availability and fee arrangements. Call Tony Colao at (800) 4-LECTUR.

OTHER MASTERMEDIA BOOKS

To order MasterMedia books, either go to your local bookstore or call (800) 334-8232.

THE PREGNANCY AND MOTHERHOOD DIARY: Planning the First Year of Your Second Career, by Susan Schiffer Stautberg, is the first and only undated appointment diary that shows how to manage pregnancy and career. ($12.95 spiralbound)

CITIES OF OPPORTUNITY: Finding the Best Place to Work, Live and Prosper in the 1990's and Beyond, by Dr. John Tepper Marlin, explores the job and living options for the next decade and into the next century. This consumer guide and handbook, written by one of the world's experts on cities, selects and features forty-six American cities and metropolitan areas. ($13.95 paper, $24.95 cloth)

THE DOLLARS AND SENSE OF DIVORCE, by Dr. Judith Briles, is the first book to combine practical tips on overcoming the legal hurdles by planning finances before, during, and after divorce. ($10.95 paper)

OUT THE ORGANIZATION: New Career Opportunities for the 1990's, by Robert and Madeleine Swain, is written for the millions of Americans whose jobs are no longer safe, whose companies are not loyal, and who face futures of uncertainty. It gives advice on finding a new job or starting your own business. ($12.95 paper)

AGING PARENTS AND YOU: A Complete Handbook to Help You Help Your Elders Maintain a Healthy, Productive and Independent Life, by Eugenia Anderson-Ellis, is a complete guide to providing care to aging relatives. It gives practical advice and resources to the adults who are helping their elders lead productive and independent lives. Revised and updated. ($9.95 paper)

CRITICISM IN YOUR LIFE: How to Give It, How to Take It, How to Make It Work for You, by Dr. Deborah Bright, offers practical advice, in an upbeat, readable, and realistic fashion, for turning criticism into control. Charts and diagrams guide the reader into managing criticism from bosses, spouses, children, friends, neighbors, in-laws, and business relations. ($17.95 cloth)

BEYOND SUCCESS: How Volunteer Service Can Help You Begin Making a Life Instead of Just a Living, by John F. Raynolds III and Eleanor Raynolds, C.B.E., is a unique how-to book targeted at business and professional people considering volunteer work, senior citizens who wish to fill leisure time meaningfully, and students trying out various career options. The book is filled with interviews with celebrities, CEOs, and average citizens who talk about the benefits of service work. ($19.95 cloth)

MANAGING IT ALL: Time-Saving Ideas for Career, Family, Relationships, and Self, by Beverly Benz Treuille and Susan Schiffer Stautberg, is written for women who are juggling careers and families. Over two hundred career women (ranging from a TV anchorwoman to an investment banker) were interviewed. The book contains many humorous anecdotes on saving time and improving the quality of life for self and family. ($9.95 paper)

YOUR HEALTHY BODY, YOUR HEALTHY LIFE: *How to Take Control of Your Medical Destiny*, by Donald B. Louria, M.D., provides precise advice and strategies that will help you to live a long and healthy life. Learn also about nutrition, exercise, vitamins, and medication, as well as how to control risk factors for major diseases. Revised and updated. ($12.95 paper)

THE CONFIDENCE FACTOR: *How Self-Esteem Can Change Your Life*, by Dr. Judith Briles, is based on a nationwide survey of six thousand men and women. Briles explores why women so often feel a lack of self-confidence and have a poor opinion of themselves. She offers step-by-step advice on becoming the person you want to be. ($9.95 paper, $18.95 cloth)

THE SOLUTION TO POLLUTION: *101 Things You Can Do to Clean Up Your Environment*, by Laurence Sombke, offers step-by-step techniques on how to conserve more energy, start a recycling center, choose biodegradable products, and even proceed with individual environmental cleanup projects. ($7.95 paper)

TAKING CONTROL OF YOUR LIFE: *The Secrets of Successful Enterprising Women*, by Gail Blanke and Kathleen Walas, is based on the authors' professional experience with Avon Products' Women of Enterprise Awards, given each year to outstanding women entrepreneurs. The authors offer a specific plan to help you gain control over your life, and include business tips and quizzes as well as beauty and lifestyle information. ($17.95 cloth)

SIDE-BY-SIDE STRATEGIES: *How Two-Career Couples Can Thrive in the Nineties*, by Jane Hershey Cuozzo and S. Diane

Graham, describes how two-career couples can learn the difference between competing with a spouse and becoming a supportive power partner. Published in hardcover as *Power Partners.* ($10.95 paper, $19.95 cloth)

DARE TO CONFRONT! *How to Intervene When Someone You Care About Has an Alcohol or Drug Problem,* by Bob Wright and Deborah George Wright, shows the reader how to use the step-by-step methods of professional interventionists to motivate drug-dependent people to accept the help they need. ($17.95 cloth)

WORK WITH ME! *How to Make the Most of Office Support Staff,* by Betsy Lazary, shows you how to find, train, and nurture the "perfect" assistant and how to best utilize your support staff professionals. ($9.95 paper)

MANN FOR ALL SEASONS: *Wit and Wisdom from* The Washington Post's *Judy Mann,* by Judy Mann, shows the columnist at her best as she writes about women, families, and the impact and politics of the women's revolution. ($9.95 paper, $19.95 cloth)

THE SOLUTION TO POLLUTION IN THE WORKPLACE, by Laurence Sombke, Terry M. Robertson and Elliot M. Kaplan, supplies employees with everything they need to know about cleaning up their workspace, including recycling, using energy efficiently, conserving water and buying recycled products and nontoxic supplies. ($9.95 paper)

THE ENVIRONMENTAL GARDENER: *The Solution to Pollution for Lawns and Gardens,* by Laurence Sombke, focuses on what each of us can do to protect our endangered plant life. A practical sourcebook and shopping guide. ($8.95 paper)

THE LOYALTY FACTOR: *Building Trust in Today's Workplace,* by Carol Kinsey Goman, Ph.D., offers techniques for restoring commitment and loyalty in the workplace. ($9.95 paper)

DARE TO CHANGE YOUR JOB—AND YOUR LIFE, by Carole Kanchier, Ph.D., provides a look at career growth and development throughout the life cycle. ($9.95 paper)

MISS AMERICA: In Pursuit of the Crown, by Ann-Marie Bivans, is an authorized guidebook to the Pageant, containing eyewitness accounts, complete historical data, and a realistic look at the trials and triumphs of the potential Miss Americas. ($19.95 paper, $27.50 cloth; b & w and color photos)

POSITIVELY OUTRAGEOUS SERVICE: New and Easy Ways to Win Customers for Life, by T. Scott Gross, identifies what the consumers of the nineties really want and how businesses can develop effective marketing strategies to answer those needs. ($14.95 paper)

BREATHING SPACE: Living and Working at a Comfortable Pace in a Sped-Up Society, by Jeff Davidson, helps readers to handle information and activity overload, and gain greater control over their lives. ($10.95 paper)

TWENTYSOMETHING: Managing and Motivating Today's New Work Force, by Lawrence J. Bradford, Ph.D., and Claire Raines, M.A., examines the work orientation of the younger generation, offering managers in businesses of all kinds a practical guide to better understand and supervise their young employees. ($22.95 cloth)

REAL LIFE 101: The Graduate's Guide to Survival, by Susan Kleinman, supplies welcome advice to those facing "real life" for the first time, focusing on work, money, health, and how to deal with freedom and responsibility. ($9.95 paper)

BALANCING ACTS! Juggling Love, Work, Family, and Recreation, by Susan Schiffer Stautberg and Marcia L. Worthing, provides strategies to achieve a balanced life by reordering priorities and setting realistic goals. ($12.95 paper)

REAL BEAUTY . . . REAL WOMEN: A Handbook for Making the Best of Your Own Good Looks, by Kathleen Walas, International Beauty and Fashion Director of Avon Products, offers expert advice on beauty and fashion to women of all ages and ethnic backgrounds. ($19.50 paper; in full color)

THE LIVING HEART BRAND NAME SHOPPER'S GUIDE, by Michael E. DeBakey, M.D., Antonio M. Gotto, Jr., M.D., D.Phil., Lynne W. Scott, M.A., R.D./L.D., and John P. Foreyt, Ph.D., lists brand-name supermarket products that are low in fat, saturated fatty acids, and cholesterol. ($12.50 paper)

MANAGING YOUR CHILD'S DIABETES, by Robert Wood Johnson IV, Sale Johnson, Casey Johnson, and Susan Kleinman, brings help to families trying to understand diabetes and control its effects. ($10.95 paper)

STEP FORWARD: Sexual Harassment in the Workplace, What You Need to Know, by Susan L. Webb, presents the facts for identifying the tell-tale signs of sexual harassment on the job, and how to deal with it. ($9.95 paper)

A TEEN'S GUIDE TO BUSINESS: The Secrets to a Successful Enterprise, by Linda Menzies, Oren S. Jenkins, and Rickell R. Fisher, provides solid information about starting your own business or working for one. ($7.95 paper)

GLORIOUS ROOTS: Recipes for Healthy, Tasty Vegetables, by Laurence Sombke, celebrates the taste, texture, and versatility of root vegetables. Contains recipes for appetizers, soups, stews, and baked, boiled, and stir-fried dishes—even desserts. ($12.95 paper)

THE OUTDOOR WOMAN: A Handbook to Adventure, by Patricia Hubbard and Stan Wass, details the lives of adventurous outdoor women and offers their ideas on how you can incorporate exciting outdoor experiences into your life. ($14.95 paper; with photos)

FLIGHT PLAN FOR LIVING: The Art of Self-Encouragement, by Patrick O'Dooley, is a life-guide organized like a pilot's flight checklist, which ensures you'll be flying "clear on top" throughout your life. ($17.95 cloth)

HOW TO GET WHAT YOU WANT FROM ALMOST ANYBODY, by T. Scott Gross, shows how to get great service, negotiate better prices, and always get what you pay for. ($9.95 paper)

TEAMBUILT: Making Teamwork Work, by Mark Sanborn, teaches business how to improve productivity, without increasing resources or expenses, by building teamwork among employers. ($19.95 cloth)

THE BIG APPLE BUSINESS AND PLEASURE GUIDE: 501 Ways to Work Smarter, Play Harder, and Live Better in New York City, by Muriel Siebert and Susan Kleinman, offers visitors and New Yorkers alike advice on how to do business in the city as well as how to enjoy its attractions. ($9.95 paper)

FINANCIAL SAVVY FOR WOMEN: A Money Book for Women of All Ages, by Dr. Judith Briles, provides a critical and in-depth look at financial structures and tools any woman wanting to achieve total independence can use. ($14.95 paper)

MIND YOUR OWN BUSINESS: And Keep It in the Family, by Marcy Syms, COO of Syms Corporation, is an effective guide for any organization, small or large, facing what is documented to be the toughest step in managing a family business—making the transition to the new generation. ($18.95 cloth)

KIDS WHO MAKE A DIFFERENCE, by Joyce M. Roché and Marie Rodriguez, is a surprising and inspiring document of some of today's toughest challenges being met—by teenagers and kids! Their courage and creativity allowed them to find practical solutions. ($8.95 paper; with photos)

ROSEY GRIER'S ALL-AMERICAN HEROS: Multicultural Success Stories, by Roosevelt "Rosey" Grier, is a wonderful collection of personal histories, told in their own words by prominent African-Americans, Latins, Asians, and native Americans; all tell of the people in their lives and choices they made in achieving public acclaim and personal success. ($9.95 paper; with portrait photos)